Silenic Drift

First Published by Strange Attractor Press 2013

Text Copyright © 2013 Strange Attractor Press / Iain Sinclair
Photographs Copyright © 2013 Anonymous Bosch

ISBN: 978-1-907222-16-0

A CIP catalogue record for this book is available from the British Library.

Strange Attractor Press
BM SAP, London, WC1N 3XX, UK
www.strangeattractor.co.uk

Printed in the United Kingdom

Silenic Drift

Iain Sinclair

Photographs by
Anonymous Bosch

SCALES

First Published by Strange Attractor Press 2013

Text Copyright © 2013 Strange Attractor Press / B. Catling
Ahnighito photographs Copyright © 2013 Rommel Pecson

ISBN: 978-1-907222-16-0

A CIP catalogue record for this book is available from the British Library.

Strange Attractor Press
BM SAP, London, WC1N 3XX, UK
www.strangeattractor.co.uk

Printed in the United Kingdom

SCALES

B. Catling

Photographs by
Rommel Pecson

SCALES is the final tale in **THE DOC QUARTET: Four forbidden Westerns with a twist in their spines** (unpublished).

Each separate story is touched, haunted or dominated by John Henry "Doc" Holliday (1851-1887), gambler, dentist, consumptive, gunfighter and the infamous friend and companion of Wyatt Earp in the legendary gunfight at Ok Corral. But this is not a tale of those famous men. In SCALES his presence is little more than a stain.

The meaning of life is that it ends.
Franz Kafka

SCALES

PILOT

FLAKES

YADILL

TENT

SCALP

HOPE

MAC

CHAINS

PILGRIMS

HELM

SIGHT

THE STONE FENCE

SAND

BLIND

BALLAST

GIN

DOWRY

PETRIFIED

PROCESSION

POLDER

FEW

BY DAWN

END

SCALES

A great stretch of land lies before us, water rushing at its edge. Twenty-seven people stand before us. Their eyes wide and clear, their faces direct. A gentle smile on each. Each holds a firearm or a farm implement. Each extends their other hand, offering a gesture of greeting. They all seem to have just arrived, seeking help and friendship, understanding and reason for being here. Behind and before them the thin strip of land is bright, in a blinding yellow landscape of sonorous wheat. The slow undulation of the crops gives the distance an hypnotic, shimmering foreverness.

Their eyes are sliding from surprise, through disappointment, into hollowness, and strain to see us as if we are on the other side of a great mist or fog. These people are ghosts; those that never made it, their mourned bones smothered by the land in-between all of their journeys. Together they weigh less than a feather. On the most accurate calibrated balances of these times they weigh nothing.

On the same scales, the *Ahnighito* - The West Greenlandic name for a tent - weighed thirty-six and a half tons. The Woman and the Dog, and probably the Man, travelled inside it with other fragments when they were locked together. And that one wrapped unity weighed almost sixty tons.

The ghosts may have travelled thousands of miles over land and sea, in rain, sun, snow and wind. To die here.

The Ahnighito and its occupants had travelled trillions of miles alone in the vast blackness of eternal cold to live here.

PILOT

Van der Linden would be broken by their consequence, but he did not know it as he stood proud in the great cracked open

land. Uncompromised slabs of mountain veered up through
the sand and cacti, to challenge the burning sky. Massive fists
of stone that seemed bewildered by their size. Some stood
alone and waiting. Some had crumbled under the primitive
winds and spiteful rains that carved and chewed them into
fantastic shapes, or smudged them to shoulder together to
create cliffs and plateaux that loomed above the vast canyon
that gashed and snaked through the visionary landscape.

Crossing it was suicidal, yet a few deluded or driven strings
of wagons tried. But never without a canyon pilot. They would
guide them and explain the consequences of every grain and
dribble of this terrible, beautiful place, although most pilgrims
never really saw it. They hurried through it with their wagons
harnessed together, their horses blindfolded and their women-
folk hiding under the possessions which rocked and rumbled
beneath the stained canvas. They said they were hiding from
the natives who owned these lands; the savages that would
kill a white man on sight. Or from the unpredictable flash
floods that would tear through the arid gullies of vermillion
sands, soaking them to blood-black in seconds.

But really they were all hiding from the openness of
this place. The desolation of total roofless exposure. All the
mountains were flat-topped, as if their sensible pinnacles had
been sliced evenly away. Scalped against optimism. A great
compress was created by it in the sweltering desert. No place
of refuge was allowed. No filter existed between the Great
Spirit and the burning earth where most men withered,
except the leather skinned Indians and the Dutchman Van der
Linden. He was an averaged sized man, who had nothing of
the average about him. His deeply lined skin was stretched
over a taut, lean body that was ageless and pondering in a way
that suggests wisdom in the still spaces between his sparing,

essential movements. He didn't waste energy or words in a landscape which was totally unimpressed by both. So he watched and balanced his curiosity about wilderness and men on every journey he made back and forth.

He had been in the canyon for years. 'Washed up' as he called it. He also called himself the 'Charon of the Vilderness'. It was his job, he ferried the travellers from one side of the gigantic valley to the other. Led their blind horses into the barrenness of absolute indifference. He even showed the scales of thickened skin that were growing from his eyelids as proof that sooner or later he would no longer need his crudely handcrafted spectacles, made of thick English pennies with slits in, that he wore against the blinding glare. Sometimes they got so hot that he had to splash them with water from his skin bag and put them back onto his face while the split head of Queen Victoria was still steaming.

It took five to seven days to get the wagons across, and two for him to ride back. He had held the job longer than any other man. They said it was because the Indians respected him. But it was fear that kept them at bay. Except that last time.

He was attacked and found mangled on the return of his last journey on the furnace road. It was said that they were Havasupai: not wild Apache or renegade scavengers, but an ancient tribe who did not fear the great loneliness, and were famed for their non-bloodthirsty wisdom. What happened on that day is unknown even to the victim. The hole in his body and the gap in his mind had been cauterized by the sun and the wind, and whatever pumped or leaked from them had vanished on both sides of the realities.

FLAKES

Robert E. Peary liked cameras. They could change him and give the hungry world the man he wanted them to see. Over the years his portraits show a heavily whiskered man full of pride, and what he thought was a kind humanity. Or a stiff backed officer of irrefutable commanding authority. Or a man clad in Inuit fur conquering the bleak Arctic. Or a family man, his love for them performed in his moving and compelling eyes. In reality, a great melancholy is haloed around his face in every exposure. There is a detachment in the cast of his gaze that never quite meets the viewer's eyes. At their best the portraits show a gruff attempt at amiability. At their worst they a reveal a hollow man made strong by determination and ruthlessness. Many of those who shared his ships, his ice and his life would confirm that all those traits were alive and shifting in the complex man, and that the balance of the contradictions was as unpredictable and as troubled as the seas and the landscapes that he chose to inhabit.

Peary was determined on leaving his mark on the world, to nail his name and life down against all the tides of forgetfulness that consumed those who went before. He achieved this in the latter part of his life by erecting the stars and stripes over 'his' spot, that he deemed to be the Geographic North Pole. Years after his distinguished life ended, the validity of his claim on the Pole came under serious question and a shadow fell across the brightness of his reputation.

But earlier in his career he sought a less conceptual prize to label his own. A trophy of pragmatic existence. Rumours of Inuits in the Savissivik cape of Greenland with iron tipped tools had been filtering back from previous Arctic expeditions. In 1818 John Ross had encountered such tribes in the North West of Greenland, while he was seeking the Northwest

Passage. He had marvelled at the ingenuity of their harpoons and flensing knives, which were bladed with flakes and shards of razor-sharp iron. This was especially significant as there was not the slightest trace of any ferric deposits in Greenland. A meteorite was the only possibility, and the quest had begun. The importance and the meaning of such a significant find had grown in the scientific imagination, and others had heard about it and its approximate whereabouts. The Norsemen and the English had been staking claims and sniffing for signs of the magnetic whereabouts. Five expeditions between 1818 and 1883 combed the snow and the tundra, and the hearts of the tribes, dispatched from the King of Denmark. From England came Rear Admiral John Franklin, and later the Arctic Expedition of 1875-76. From Sweden came Baron Nordenskiold. But all failed to find the source of the iron until Peary bribed a local guide with a gun and a promise of friendship. At Saviksoah Island the trail got warm. The Iron Mountain was in the American's sights.

YADILL

Wounds are funny things. If you have never had one, then you have always lived alone. Van der Linden's wound took two months to heal, and six before it started prophesying. It did this through a series of aches and pains; the mildest predicting changes in weather, and the staunchest warning of bloodshed or attack. The wound kept him alive longer than anybody else who dared to take employment from the canyon, but it had not seemed so in those early days when he was being doctored at McAlister's.

And then became it. The infamous trading post had its own water and was the only manmade structure in the wilderness; three collapsing rooms under a slouched roof on the lip of nothingness.

"For fuck's sake shut him up, or I will!" bellowed Tully, who was sitting in the artificial gloom of the trading station. Outside it was just beyond noon, and 110 degrees with a wind that had picked up every speck of sand and grit and held it up to the sun for inspection. This abrasive haze bucked and hammered at the shuttered windows and bolted doors, and bellowed down the chimney and into every whistling crevice with a vindictive joy that had showed no fatigue. Some of the gusts were powerful enough to knock a horse over, and everybody knew that its secret ambition was to sandblast another man down to the pinks and whites of his bones. It had done it before and it would do it again. They had been holed up inside for three days now. Tully was trying to stay drunk, but the incentive was wearing thin on the few last dollars he had stashed under his gun belt. The noise of the injured stranger in the back room was stealing his inebriation and costing him more than he had. Tully was not the kind of man to be stolen from. His temper was sulphurous and he had scant respect for the laws - and the rattling hutch in the desert was a long ways from them.

The owner who everybody passing through called McAlister, wasn't McAlister. His real name was Turquint and he stood behind his bar like a captain of a constant and durable vessel, which indeed it was. McAlister's, in one manifestation or the other, had held its ground against savages, sandstorms and the extremes of temperature for more than twenty years. An astonishing amount of time. Ghost towns and the bitter rinds of long devoured dwellings and long forgotten lives littered the safe sides of the canyon for miles and decades.

McAlister's had sheltered many and exchanged their belongings, lives and possessions for tough money. It also saved the lives of lost pilgrims and exhausted pioneers. It absorbed the lame, the lost, the wicked and the wild and forced them to pay dearly for its sanctuary. A good cross-section of those characters now sat or lay in its sweltering thick shadows while the wind pried and gibbered through every gap and fissure, especially in the back room, which was the oldest part of the structure. Even with the shutters tightly closed, rods of light hammered through the splintered holes and cracks. Smaller threads of vicious light spidered the darkness where the bloated oil lamp bobbed about above the restrained man. They dared not open the shutters. His wound could be full of sand in seconds.

The Dutchman's screams added a counterpoint to the symphonic moans and gusting rattles. The hole that the Barber and the Hopi were trying to close was torn between his chest, just below the collarbone. and his armpit on the left hand side. It was impossible to tell which was the exit and entrance of the wound, or what the hell had made it. The Barber's best guess was a spear, but the Hopi shook his head in disagreement. Whatever had pierced him had missed the vital organs and major vessels, but its bent glancing blow had skewered and skinned some of the nerve web of the Brachial Plexus. As they swabbed the wound and sewed its ragged edges, the patient jumped and bucked against the restraining ropes that the Barber feared might cause more damage. Most of the blood had stopped flowing and the Barber was amazed by the size of the cavity as he tried to keep sand out of its sticky interior. The wet ball of whiskey-soaked old newspaper that he was using for staunching the wound was already dimly sparkling. The swaying yellow light of the oil lamp that the

Hopi help aloft caught the mineral glint of each hard grain against the blackening blood.

"This hole is far too big. Looks like he's lost his tripe or a second heart somewheres."

The Hopi peered inside and nodded without understanding most of what the Barber said, but he was now saying it loud enough for those drinking in the next room to hear.

"It's too damn empty. He's lost a chunk of himself somewheres, but for the life of me I don't know what it was."

He grabbed at the Hopi's arm and pulled the light in closer. The sutures were holding. There was nothing he could do about the mangled nerves, and he thought it best to close him up. Then he saw the mark. He looked closer and saw that it formed a word. YADILL. His jaw dropped in amazement as the mystery intensified.

"I'll be damned."

He blindly reached for the lamp again to pull it even closer. His fingertips touched and stuck for a moment on the hot glass until he sharply flinched "Fuck", and then bellowed at the Indian, "BRING IT CLOSER". The smoking lamp was now practically in the wound.

Turquint appeared through the back of the bar and occupied a position where he could see both rooms of possible trouble at once. Some unkind folk said that was why his eyes were like that, poached on each side of his head and looking in different directions. Kinder and more nervous folk explained that he was born like that, and it was why he was so good with a sawn-off double-barrelled shotgun. Generally the dialogue about his goggle eyes stopped there.

"Yadill... it says YADILL."

The word was printed on the pink open muscle. Some of

the letters were back to front, but it was clear enough.

"I'll be damned. Never seen anything like it. Someone has taken flesh and replaced it with a word, kinda signed it. I'll be Goddamned. The strangest thing I've ever seen."

The Barber was exhilarated and light-headed. Wonder had combined with the stress, concentration and the amount of raw whiskey they had been using, mainly as an antiseptic. He stepped back from the groaning Dutchman and sat down vacantly on a low stool.

"I'll be damned."

He had used Yadril. He was racking his brains to remember the patients. Its proprietary name was trimethanol-allyl-carbide, and it was used as a vigorous antiseptic. He could have used some now. Is that what this was about? This man's body miraculously printing out its own prescription, giving treatment suggestions to its surgeon? But surely that was nonsense. He took a long deep slug of the whiskey to wash his foolishness away. The Hopi lifted the lamp and looked for himself, turning his head this way and that like a curious cat. Then he grinned solemnly and addressed the whistling room.

"From Papa," he said.

The Barber did not hear it at first.

"From Papa, it come from Papa."

"Whose Papa?" the Barber eventually asked, inadvertently catching one of Turquint's eyes. The proprietor shrugged indifferently, as if to dismiss or deny that it might be his.

"Whose Papa?" the Barber asked again.

"Yous." The Indian pointed at him.

"My Papa inside him?"

One of the drunks from the next room was suddenly hanging in the frame of the door, his bleary eyes trying to see into the dim room to locate the source and the meaning of the commotion. The Indian pointed again, but this time his aim was to the right of the accused man.

"You mean HIS Papa, don't ya? His name kinda tattooed inside..."

The Barber's voice was irritated and exhausted, but the slant of its hot annoyance was cocked to speak again, when the Indian stepped forward, his pointing finger outstretched and slowly moving across the room, hooking and pulling seven staring eyes towards its absolute target. The finger stopped when it touched the forgotten sticky ball in the Barber's hand.

"Papa," the Indian said.

The frowning Barber stared at him and then at his hand, then slowly stood up and returned to the Dutchman who was now quiet. He carefully took the lamp and stood over the wound. Then he looked for a long time at the stiffening ball of paper in his other hand. He tried to let it go, but it had adhered to his palm and the insides of his fingers. A blank rage suddenly overcame him and he violently clawed the thing away with his other hand and threw it into a far corner of the room. He grabbed the medicinal whiskey bottle and doused his hand, wiping it on his trousers.

"Bring the light," he commanded and peered back into the unconscious man's wound. He then looked about frantically. Finally, it seemed in a kind of desperation, he started tugging at his own shirttails, pulling them out and ripping a section of the worn material away. He soaked it in whiskey and thrust his hand back inside the Dutchman, daubing then scrubbing at the word. The wounded man screamed awake and the shock of it made the Barber aware of what he was doing. It also made Turquint automatically touch the stock of one of his shotguns under the bar, and the drunk shivered in the door frame. The Barber was horrified as he looked at the stained rag and back to the pink muscle that was bleeding again, and now only had a bluish stained blur where the word had been before. Like a somnambulist he looked into the oily yellow light of the lamp and whispered to it instead of the Indian.

"Close him up. Better close him up."

The Hopi seemed very pleased to do so, and started to collect the thread and curved needles.

"Wassgoinun?" the man at the door asked before he slithered down the frame and onto the sawdust floor, where he lay beyond the screams of the wind, the sand and the Dutchman.

TENT

Peary's decision to take his wife and daughter up into the Arctic waste was again greatly frowned upon. The frozen seas of Greenland were considered unsuitable for the fairer sex and impossible for a babe in arms. No one knew if he did it out of perversity or sentiment, but both were deemed irresponsible. Marie, his Snow Child as the Inuits named her, had been born in Greenland and her parents quarrelled into agreement that she would live equally between there and the States. She had been second-named Ahnighito, after the Tent of iron. She would be there when he finally captured it and brought it back to Manhattan on his third attempt. This time it really would be his. He had packed the *Hope* with bigger hydraulic jacks and packed her gunwales with massive beams of sturdy white oak and gneissose boulders to increase her ballast. Every inch of the ship was packed with the mechanisms and materials of his resolve, and at its core was the deconstructed blank of the carriage that would contain the celestial iron mass. The coal for the ship's engine weighed less than his cargo of rails of cast iron and steel that he had brought to shift the prize. He had gathered his trusty captain and crew, then sprinkled his delicate family in their cabins and steamed North.

On his first expedition he had gathered the Woman and the Dog and un-snowed the Tent, digging deep into the frozen roots that wanted to clench the vast iron core for its own. The Eskimos knew that this land did not own it. They told tales of its fiery arrival, crushing the earth. Stories passed down the centuries from a time before human beings existed. It had broken on impact into the Tent, Dog and Woman, and the Man that they all knew was also there somewhere, hiding under the cold and away from their chipping tools.

Over all their generations they had attacked the hardness of the smaller demon stones relentlessly, gathering the scales of iron to tip their spears and arrows. That is how Peary found it. He trod in the prints of rumour to the Dog and the Woman, then stalked the invisibility of the buried greater until it took him to the spot where his compasses bucked and gibbered. He and his men and the recruited natives started digging until their spades became warm. They worked ceaselessly for days, only stopping to gain breath and smell the air for the advancing ice, from the thin pale light of the morning until the blue evening overcame them, turning to purple under the pale surface of all things. Sometimes their tools sparked light from it; an icy plume of fire that put the whiteness of snow to shame. Each night they retired in triumph and slept poorly, waiting for the next day.

Gradually its dark bronze contour was becoming clear, an irregular wedge rising up like a great dorsal fin.

By the afternoon of the fourth day they had it dug clear, but still without knowing how deeply it was sunk into the now bare permafrost. It stood and looked at them in surprise; solid iron, brown and unlike anything else around them, or in that land, or on their planet. Dull and indifferent to this world, its petty physics and the creatures that scrabbled around it. Peary told them that tomorrow they would move it towards the boat. Nobody believed him.

The next day they brought their biggest screw jacks and buttressed them against planking laid across the firmest snow. They tightened the ten-tonner until the snow screeched, the wood splintered and the steel jacks mangled under the effort. The dull brown mass of solid celestial iron, which vaguely smelt of blood, had not moved an inch. The men laid their tools aside and reassessed its amazing brutal power of resistance.

The Tent glared back at them, implacable and defiant. Peary knew that all their tools were too insubstantial to impress the 'Iron mountain', as indeed was the *Kite*. Her dainty lines were already holding their slender breath against the weight of the Woman and Dog. The ship might just fold up, like so much kindling under the weight of this monster. Finally he thought about how they had ferried the other irons to the ship on a great raft of thick sea ice. The ice had bobbed and yawed uncontrollable, threatening to pitch the treasures into the depth of the bay, while the gantries of the *Kite* screamed and crunched, lifting the meteorites on board. The ship was violently listing and the men who were not engaged with the loading ran to her rising side to give extra ballast there and keep her gleaming keel below the tightening surface of the freezing sea.

It was impossible. He had been defeated again, cheated of what he had planned to be and for a few moments, while his men waited to be told what happened next, he was that pale crushed child again, returning home to his widowed mother beaten and bruised, the little girl's sun hat that she made him wear torn beyond repair but still tight in his hands. Finally he gathered himself and he gave the command to leave the tools.

"We will return next year with a bigger ship," he said in a voice that frightened the child.

SCALP

One morning the Dutchman awoke with words in his mouth. The storm had been over for two days but Turquint, called McAlister, was still sweeping sand away from the door. Further along the outside it buttressed the wall in smooth, irregular dunes. The door was wide open and the heat and the brightness of the day flooded and tried to clean the gloom and the stink of men. Only McAlister and the Hopi were there.

"Waar ben I?"

The Indian heard him and hobbled to the door to drag the owner back inside.

"My God it speaks!"

He was wiping his hands on his apron and his eyes wobbled like wet poached eggs as they adjusted to the talking shadowed blur. Van der Linden was looking around the filthy room.

"Where am I?"

"Yo bin sick Mister. All cut up by savages."

The man's voice seemed detached from his face and the wounded man tried to follow it around the room. The Hopi brought him water in the weightless shell of a dried out vegetable gourd. He still could do very little by himself, and the Indian helped him drink. While his mouth was full of water McAlister spoke again.

"Chava, here, found you out in the canyon and brought you back."

The Dutchman twisted around to look at the Hopi.

"Dank you," he winced.

The Indian said nothing and looked blank.

"He found you out near his traps, you and your horse. How come they didn't take your horse?"

"Don't know about the horse," he said, still trying to understand the Hopi's face.

"Now, one thing I gotta tell you, before you're strong enough to get angry is that you ain't got no possessions anymore. No horse. Nothing." As he spoke, McAlister was back behind the bar where his guns lived. "Do you understand me and all?"

The wounded man nodded. He had no memory of ever owning a thing.

"The Barber took it all for doctoring you. He saved your life and that's for sure. And now he's gone."

In the thinking over of all this, the wounded man's eyes crossed and strangled his speech until he fainted again.

"Hout coll," said the Hopi and McAlister sniffed and sidled into the other empty bar.

All the McAlisters had survived running this dangerous shit-hole by controlling the invisibility between their total absence and absolute dominance. Some had used will, and others instinct. All of them had to know the way of men, especially the bad ones. Good ones never found their way to this establishment.

Three days later, over corn mash and prairie hare, McAlister said,

"Yous and me gotta talk about how you're gonna pay your keep."

"Ok."

"Cos I bin feedin' ya for a good while now, and it looks like you ain't going nowheres too soon."

"Ok."

"Well, how ya goin' to do that?"

"Ok. Soon, when the next tallyman comes in."

McAlister looked at the Hopi in desperation for any sign of understanding or sympathy. He looked back with eyes that were almost totally transparent. The owner turned back to the wounded man and his chow.

"Is your name Yadill?" he said, his mouth full of mash.

"No."

"Waddisit?"

"Jacobus Van der Linden."

"That's what I thought. You been here before, runnin' the canyons ain't it?"

"I have sometimes taken wagon trains from here."

McAlister frowned and forked more food into his speech.

"German?"

"I am a Hollander from the Brabant."

"Icallyadutchok?"

"Ok."

Two days later Tully came back. He was in a worse mood than before, sitting in the back of the first room, drinking hard. The wounded man heard his abusive tone and the cursing and shouting, and in someway it was attached to his pain. The sound of it stitched into his butchered shoulder. He got up and staggered through the doorway into the bar. Tully could barely be bothered to see him.

"I am Van der Linden, with a hole inside him. How much you want for your scalp?"

It took a long while for Tully to understand the words and then even longer to believe that somebody had dared to say them, especially the tottering human wreckage in the thin doorway. He got to his feet and craned his neck to look even closer at the skeleton covered in bandages.

"What are you, a jipshum mummy?"

"I need pay to patch up my hole," Van der Linden said faintly.

Tully laughed and hooked his thumb into his gun belt.

"I'll give ya the toe of my boot to patch up your hole," he squawked with the joy of the cleverness of his wit. He reminded himself to tell others of it. Van der Linden drained of colour and slid down the door frame to the floor with a yelp of pain. He then slithered over. None of the men present had ever seen anything like it. A human snake. McAlister was spooked and came out from behind the other end of the bar.

"Am I supposed to be scared?" guffawed Tully.

"You should be," said the Hopi suddenly.

"Fuck off dog meat. No one's talking to no dog meat," said Tully, pulling a heavy hunting knife from behind his back. "Perhaps we better skin it?"

McAlister was between them and missing the sanctum of his bar. One of his eyes was still on it, the other on Tully.

"Let's all have a drink and calm down," he said.

"No one tells me to calm down or calm up," Tully said, suddenly grabbing the owner's shirtfront. It ripped loudly and buttons spewed across the floor. The curve of the over-sharpened blade was under the owner's neck.

"I've spent a King's ransom in here drinking your watered down piss. You ain't gonna tell me nothing."

McAlister opened his mouth and his jaw seemed to make a tense metallic click, which surprised all of them, until they heard the knocking from the bar. One of the proprietor's double-barrelled shotguns had been cocked and was slowly tapping its barrels on the cankered surface. Behind it stood Van der Linden.

"Scalp him," he said.

Tully's hand slid down towards a six-gun that sat on his hip, the holster tied down to his thigh. The shotgun was up and moving its barrel pointing into Tully's face all the way as it crossed the room. By the time he got within spitting distance Tully's hand had formed the negative shape of the protruding butt of his revolver. The knife was still at McAlister's throat.

"Ok," the Dutchman said and casually lowered the shotgun, locking his sad, resigned eyes on the gaze of Tully. They were defeated and appeared to have lost their moment of muddied resolve. Tully was trying to think of a reason not to shoot this pathetic weakling when the shotgun fired at point blank range and blew all the poised fingers and their twitching clasp into mush. It also removed the horn handle of Tully's 45 and a considerable chunk of his waist. The knife spun sideways as he was blown backwards, screaming.

"Now scalp him," said Van der Linden, swivelling the other barrel into McAlister's face.

"I can't do that," he shouted loud enough to frighten a dog. His eyes bulged in different directions.

"Do it or I will shoot you in half and get him to scalp you."

The Hopi did not like being in the equation, but saw the funny side of it anyway and started laughing.

"Ha ha ha."

Tully was sprawled across the floor and had pulled the butt-less gun out of its shredded holster with his good hand. It was useless, the hammer spring had gone with the grips and their skeletal frame. It slid about in his hand in what looked like its own blood like a just gutted, slippery fish. The Dutchman stood over him, waving the shotgun.

"Ha ha ha," laughed the Hopi.

"Sit quiet now while you get tonsured."

"Fuck you, I'm gonna cut your fucking heart out, fucking cunt."

"Ha ha ha."

"Hold him down" the Dutchman said to the Indian, who stopped laughing and obeyed.

"It will hurt less if you are still and he is quick."

He waved the shotgun about like a maestro, conducting the act of retrieving the knife and doing the slicing. Tully did not take the good advice and kicked like a drowning mule. But the Hopi held him tight and McAlister sawed away. Two dogs would have been frightened by the noise he made.

When it was over Tully sank back to the floor, his hand clamped over the raw meat of his head. McAlister was bone white. He stood with the knife in one hand and the wet scalp in the other, holding it out, trembling towards the Dutchman.

"Thank you, but no, it is for you."

All eyes were on the scalp.

"Me?" said the owner.

"Yes. It's the price of your hospitality, paid."

After a while the Indian started laughing again.

"Ha ha ha."

They opened another bottle and waited for Tully's head to coagulate. They had picked him up and propped him against the back wall in his usual drinking spot. McAlister had bandaged the misshapen stump of his hand that still had a little finger intact and working. He had also given him a rag to tie around his head and an old flour sack to patch up the bleeding of his hip. Dutch sat on the other side of the bar with both of McAlister's loaded shotguns on the table before him. He had told the owner to line the bottles up in front of Tully. He wanted him helplessly drunk for at least a few days. He also knew that the goggle-eyed owner would attack him whenever given the opportunity. He knew that the moment McAlister laid his hands on another of his secreted weapons he would be shot in the back and thrown back into the canyon for the coyotes. So normal sleep was out of the question, and he had to find a way to tether the man. The Indian was another matter. He was so inscrutable that it was impossible to know which way he would jump, if at all. So he decided to ask him.

"Chava, is that right? Chava is your name?"

The Hopi turned in his direction and said very slowly.

"C h a v a t a n g a k w u n a."

Dutch stared, trying to understand the sounds through his eyes.

"My name is C h a v a t a n g a k w u n a."

"Your name is long like mine."

"It means short rainbow."

Dutch looked at the weathered man who lived somewhere between this rank shed and the wild prairie and wondered what kind of child he had been to be given a name like that.

"You Yadill from the Papa."

"You can call me Dutch. I don't know about Yadill or his Papa."

"Papa still here, him never go away. Sweep only sand here, maybe burn, but not sweep."

He said these words very seriously and Dutch nodded out of assumed comradeship without understanding a word.

"You want me find?"

The Dutchman made a series of noises and gestures that smudged all decision while showing some enthusiasm in answering. The Hopi could make up his own mind about the answer to his impenetrable question.

Tully let out a guttural groan, fell sideways and vomited.

"How much longer we gotta keep him like this?" asked McAlister.

"Until we all become friends again, including you."

"Jeez, you got a fucking strange way of making friends."

"Shall I trust you?"

"Why not? I am a good man."

The Dutchman stood up, one shotgun in each hand.

"Choose," he said to the owner.

McAlister reached out and gripped one of his beloved guns.

"Is it loaded?"

"Both barrels."

He took the gun and walked a few paces away.

"You've got a choice, you can wound yourself like me or fight to the death."

"You're crazy."

"It's the only way to trust you."

McAlister's eyes tightened. He took another step back.

"You really want this?" he said looking into the Dutchman's eyes. Cold restrained seas were held back there. It was the wound that was talking, and it was ready to construct another polder. McAlister turned the gun slowly and rested the mouth of the barrels just under his collarbone. He closed his eyes and squeezed the trigger. It clicked empty.

"Danke," said Dutch and fired both barrels into the previous owner.

The Indian looked at him.

"His wound would never forgive me and he knew that. He was playing for time. I don't have that long."

The Indian nodded at his wisdom.

"Do you want to be the McAlister?" Dutch asked him.

"Never."

"Then I guess it's me."

Chava buried Turquint in the warm arid dirt at the back of the establishment. Wooden stumps of what use to be crucifixes stuck out of the ground like broken teeth. They were many. The hallowed resting places of the past McAlisters. Maybe even the first, the one who found the water and dug the well.

Dutch came out to watch as he patted the ground flat with an out of tune spade.

"Where's Tully?" he asked.

"Under here with him."

"Was he dead?"

The Indian turned to look at him resting on the spade and wiping sweat from his face.

"Nearly," he said.

HOPE

In the second expedition the newly acquired steam vessel the *Hope* brought heavier jacks and lifting tackle, but the ice was not giving way. There was only a small lane of open water when they entered Saviksoah Bay at noon on the 22nd of August 1896. A barrier of ice-pans slowed them in reaching the natural pier below Meteorite Island. But Peary and his men paid no attention to that as they moved to the starboard side of the docking vessel, staring and pointing at the dark bronze crest of the Ahnighito peering out from the debris of last year's mission. A cheer rose up and even before the ropes were fully tied men had jumped ashore, shovels and pickaxes in their eager hands. By supper time the brown monster stood clear of the snow and rubble around it and from that moment, for the next ten days, day and night, they tried to prise it free and roll or drag it closer to the *Hope*.

The Eskimos were greatly impressed by the weight and mechanical robustness of the jacks and believed that 'Saviksoah' (the great iron) had finally met its equal. Steel tracks were laid and two huge steel chains were ready to haul it down hill towards the sea and the winching huts. It was only when the two thirty-two ton jacks collapsed and gave up the ghost that a genuine suspicion arose like a tangible miasma around the 'iron mountain' that it might indeed possess some inherent supernatural property. No one present had ever witnessed such obstinacy. Everything in its path of controlled toppling was crushed utterly. The very ground itself would dent and fracture under its unpredictable irregularity. It offered no plane or surface for attachment. Every angle seemed designed to make the chains, steel cables and metal bars slide off. It was a deity of inertia, put into existence so that men or creatures on other planets may exhaust their muscle, will and knowledge on it and give up their worn out blood. It even fatigued logic and imagination, because its form could not be described. From every angle of its turning it became a different shape.

Peary later wrote:

"The inherent devilry of inanimate objects was never more strikingly illustrated than in this monster."

Then the sixty-ton jack broke down.

He only had the unwieldy hundred-ton jack left by the time they reached the flat ground at the base of the hill. The last bit of gravity he had on his side had fled and the ice was waiting for no man or meteorite. On the last night a fierce Arctic storm blew in. It rattled the snow-covered ship and wove its moaning voice in the rigging. The sea outside was churning a murky green under the yellow, full moonlight. A vast halo had grown around it, scything the flying clouds. Further out in the bay the crushed icebergs were becoming dislodged and the wall of protection was quickly breaking up. Peary was writing in his journal and preparing for the disappointed journey home. The snug warmth of the cabin gave great comfort as the ship was jostled and played with. Albert Operti made another lunge for the door, letting in blasts of cold air and horizontal snow.

"For God's sake man, make up your mind. IN or OUT.."

Peary spat at the man as he disappeared back into the storm. The words were not heard or understood. Peary developed a violent lisp when he was enraged. And much of what he said then could not be comprehended, especially when it became entangled and muffled in his vast over groomed walrus moustache. Sometimes the sight and sound of his lisping voice activating the appendage produced a bemused expression on the face of the listener. Any sign of a grin or a twitch of a smile increased Peary's rage, so that the next sentence was even more deformed than the last, thus creating a horrible automatic spiral which would lead to one of the Commander's intolerable furies which were mostly finished with an act of spiteful cruelty.

This was Operti's fourth excursion outside in a two hour period. When he came back to the cabin to get warm and dry he was restless and kept pressing his face against the porthole window, muttering "magnifico" and "extraordinary." He also kept trying to engage Peary in conversation about visual Arctic phenomena. Artists were valuable members of most expeditions, but often the price one had to pay for their talents of recording was dealing with their obsessive eccentricities. This one had recently completed a tolerable portrait of his leader and was thus under grudging acceptance.

"Have you ever seen a moon bow like this before?" Operti asked.

"I haven't seen this one yet, I have been in here trying to get on with this."

The artist did not receive the hint.

"BUT YOU MUST! You must, it is, it is…"

"Magfnifficfent?" suggested Peary.

Operti looked into the shape of the puzzling sound on the tight face and then understood.

"Si, magnifico. Yes."

He then quickly gathered his sketchbooks and paint satchel and rushed at the door again, deaf to Peary's continuing complaint. After Operti had gone, he bolted the door and shuddered in the remnants of the icy blast. He stabbed at the tiny coal fire with a metal poker until it obeyed and burst into flame.

He had written for ten minutes when the sound came. At first he only heard it deep in his soul, then his ears alerted him to the strangeness. He did not even have time for annoyance. A great dark sonorous bell was echoing out into the storm. A sound of such profound resonance that even the wind seemed to hold its breath. He was out of his chair, into his Arctic furs and out of the door before the next toll sounded. The deck was slippery with snow and the rigging was thrashing alive. He looked around at the solid sea and the fleeting sky, and then the moon, as his eyes touched the vivid corona at a far distance. Then the bell rang again, perfectly resounding the halo in the sky. Outside it was louder and overpoweringly mournful, but touched with an animal savagery. His eyes were still fixed on the lunar ring, and for a profound second or two he was convinced that the sound emanated from it, or was a manifestation of its solemn unearthly power. He was speechless and totally in awe, when Operti appeared at his side and said,

"Magnificent, yes."

Before he had time to react to the intrusion it sounded again, but this time it was not from above. Peary had turned his great fur-lined hood to confront the artist and the sound had peeled at him sideways. It was coming from the shore, from the jetty. He strained his eyes towards the small red glare coming from the Eskimos' evening campfire close to the mass of the iron mountain. Then it sounded out again and he understood everything. He ignored the artist and skidded across the deck to the gangplank, which twisted treacherously in the rising and falling waves. He grabbed the ropes and hurried across to where his men stood, one with a heavy wooden pole in his hands. He had a respect for the Inuits. They did what they were told and they trusted him with great admiration, which in fact was fear. Nevertheless, they had showed how to survive the Arctic extremes. He wore their clothes and ate their food when pushed. They had saved some of his toes from frostbite and he had shared the warmth of their women and sired offspring by them. 'Little Brown Wizards' and 'little fur-clad children of the ice flows' he called them. He had listened to them and heard what he wanted to hear. He always wanted them to work with him. 'Peary System' he called it and in the snow, under his command, they could be very efficient.

Seeing him stand loomingly in their presence, all notion of them being 'children' disappeared. They were hard-jawed and rigid about their task. Only their height suggested anything neotenous about them. But Peary should have known better. He knew what strength existed in their compacted bodies, what rigour was bound in the denseness of their claddings of fat and how the form of it was always adult and sensual. If he had not gleaned that from his hunting, then he should have known it from his bedroom. He had had his young Greenlandic 'wife', Alakahsingwah, photographed naked more than once. The pictures showed her solid lithe power, and he had showed her and her pictures to many, in pride of his ownership of her strength and barbaric comeliness.

The men were never treacherous, that is why he listened to them - even to the husband of Alakahsingwah, who worked here for him and never once complained about the use of his wife. Peary listened to everything, because even inside their fairy tales was a hard kernel of knowledge. Hugh Lee had translated the fears of the Inuk Panikpah and Tallekoteah, explaining that the evil spirits had cast the Woman and her Dog down from the sky wrapped in the volume of the Tent, the Ahnighito. He also said that they chipped away at the smaller irons, but feared to touch the larger because it still contained demonic properties. The truth of this existed in the memories of those people. The lie of it was invented to appease the invaders.

Peary pondered over the contradiction of what he had been told and what he was now witnessing. He turned to Tornaarsuk, the man with the wooden baton.

"Why do you strike the demon Ahnighito without fear?"

One of the other men translated the question. Tornaarsuk thought long and hard and explained.

"It is now at the lip of land and sea and wants to return back to its bed. To hide again beneath the snow and the years. So I beat it to get its voice caught on the Qaammatip qinngornera. The wheel around the moon. Snared there. Now it cannot go and hide itself again."

At the end of his speech he stepped forward and handed Peary the club, who saw that Tornaarsuk was telling the truth. The meaning of this tale showed great dedication and loyalty. He took the club and was just about to strike the mass himself when he saw something else. He stepped closer to the block and stared in amazement. The wind had dropped slightly and the snow was not as horizontal as before. It swirled around them and fell vertically, settling on them and everything else, except the Ahnighito.

All the men followed their leader and watched in amazement as the large flakes refused to settle on its implacable surface. The moment they touched it, they

melted away, as if some residue of the celestial friction that had once made it glow like the sun was still burning in its dense, solid innards, so that nothing was allowed to lay hand or atmosphere on the surface of its strangeness. Even the red fires only reflected as bruised smudges on its resistant enigma. Peary quietly put the stick down as the artist came to his side.

"Do you see it?" he whispered in awe.

"Oh yes,' he said chirpily.

"It's very brown."

Albert Operti was not invited on the third expedition to retrieve the Ahnighito. Peary's wife, Josephine, and the Snow Child where there to give him the support and luck he needed and to show his 'Brown Children' that he had no fear of the demonic iron. His only apprehension came as they

drew near their final destination. He dreaded that it would
be gone; stolen by another piratical nation. Or worse, that it
had suffered a landslide or an iceberg collision and had been
avalanched into the freezing sea to a depth that he could
never reach. Or worse further, as in his nightmares - where
the *Hope* ploughed through the complaining ice that shivered
apart and scratched at the ship's sides, trying to reach his
wooden bunk - that the night sky had sucked it back up to
repudiate his notions of petty ownership. That it had snapped
its charm of sound that tethered it to the moon's halo and
raced it through the black night, ignoring all influences of
gravity until it reached its point of origin, where it slowed and
finally locked in to become a pinprick of grinning light, from
where it would watch his life pass in an insignificant flicker
from a thousand million leagues away.

His telescope strained through the morning fog as the ship
edged and bullied its way past the drifting bergs. His compass
and his heart told him that they were near. He checked the
magnetic reading again and was about to close the instrument
when he saw it tick. He tapped the case and worked the
restraining lever. The needle quivered and tried out different
modes of alignment before settling incorrectly three degrees
off of North. He tried it again and again. There was no
doubt, something other than the poles were tugging at the
redoubtable needle. It could only be the Ahnighito calling to
him through the thick fog, its influence equally searching for
him, awaiting his return.

    An hour later the air cleared and the jagged shore floated
before them. The trembling block of the chocolate coloured
shadow slid between his lenses and made its implacable
presence finally known.

They warped *Hope* in against the compacted ice that shouldered the stone jetty. Peary never let his eyes waver from their grip on the Ahnighito. He did not even hear his wife speaking to him as she snuggled against him. Peary had other wives in these parts; younger women who kept the cold at bay and were energetic in their responses to his significant desires. Alakahsingwah was his most constant. She was only fourteen when he first touched her overpowering warmth. She had become an incentive to his returns. The musk of her unquestioning intensity wedded to his addiction of the polar wilderness. But not this time. He had brought Josephine to keep him constant, to make the link between the strangeness of the great iron and its desired resting place back in Manhattan. She would de-jinx the failing jacks and drive all doubt and weakness out of the winches. She, like the white oak, was there to conquer the demons that held the meteorite to the ground by fear and gravity.

MAC

And so it came to pass that the wilderness pilot, who had worked the furnace canyons and was born among the division of waters in a land where only trees and church spires dared raise their heads above man and high towards God, became McAlister. And in this new land his new name was attached to the ramshackle buildings that contained their own well; the only well for fifty miles that stayed wet while the arid land around withered in a great dryness, and the scarcity of water forced all languages to shrink and contain themselves like a prairie cactus. Thus Jacobus Van der Linden begat Dutch and Dutch begat McAlister, and McAlister begat Mac. Even though the name Yadill never really went away.

It was probably the Barber who perpetrated and prolonged it by telling folks the story far and wide. Chavatangakwuna did not understand the problem at first. He thought that he knew the answer and kept the proof that the new McAlister never asked to see. But after a while he began to doubt his own wisdom in the matter. He did not wish to talk about anything which might cause a problem between them, because they got on so well. This was the first white man he had ever met who showed that civilisation was possible inside them. Mac even understood his refusal to touch Tully's scalp, and to nail it up over the bar himself.

"Mac, we need some whiskey out here," called Dodgem from the prowl of the bar. His two companions had already found their seats in the hot gloom, and one was pulling off his dust-filled boots. Mac came from behind the bar, a bottle in his hand.

"You fellers want eat something?" he asked.

They all looked up at the odd voice. Dodgem said it first.

"You ain't McAlister."

Mac said nothing, uncorked the bottle and set three glasses on the scarred and filthy surface of the bar.

"He had goggle eyes, like a fish."

"You want a cut?"

"And he was smaller and talked American."

"I got turpentine and molasses and pepper cordial."

"No just water, a lot of water," said the seated man with bare feet.

"You want maybe wash your feet?"

"No, just to drink."

Dodgem gave up asking questions about identity. Such things were unnecessary and downright dangerous in the badlands. He scratched his head, retrieved the whiskey and glasses and carried them over to his friends. Chava appeared from a side room with a jug of cool water.

Dodgem, Cougar and Brace stayed three days. They behaved and paid well and brought some news of the outside world. On leaving they bade good farewells and Dodgem said,

"I hope you're the same Mac next time we pass through. You cook better than the last one."

"I will be," said Mac, who had forgotten Dutch and all the land outside, and all the seas of sand and water.

CHAINS

Finally the block of iron stood facing the ship. It was held in the massive oak and steel buggy that Peary had built around it. The steel railway tracks buckled beneath it but held their line across the white oak gangway and onto the bobbing ship. This was the moment he had been waiting for. All the twisted metal bones of the broken jacks and splintered props lay jagged in the snow behind him: shipwrecked skeletons of the meteorite's short halting journey across the snow, made over the years of failed attempts.

Now, at last, it was yanked from the mouth of the land and pondered the frail jaw of the ship. All the men stood back as Peary draped the block with the Stars and Stripes. Josephine was at his side and a small bottle of wine was placed in the Snow Child's hands. She was held up towards the mass and he took her tiny mittened hands in his and drove the bottle forward. The wine sizzled and froze and the glass vanished under the snow. A cheer arose, and then the serious work began.

The winch on the ship began to pull and a strange moan

arose, while nothing moved. The sound increased and some of the Inuits stepped away. It was impossible to trace where the voice was coming from; the insides of the ship, the hard muscle of white oak or the collapsed night interior of the iron. They all seemed to be sharing or carrying a part of it. Then a movement changed its pitch. It was not the Ahnighito shifting, but the *Hope* being pulled against the dock. The ship moved sideways and creaked ferociously in its restraint. The granite-hard icicles that had formed and remained unbreakable throughout all the seasons of freezing storms were crushed or sheared off with the loudness of gunfire. The ribs of the ship were braced against their own collapse, as it was squeezed between the power of the winch and the resistance of the mighty weight.

The crewmen looked at Peary, waiting for him to give the command to stop. Instead he flew at the meteorite and pushed his puny weight against it. Then so did the rest. A shiver ran through it and it budged forward. The voice changed to a lower moan and the Hope held her ground within the sternness of her ribcage. It was moving along the track, gathering a sluggish momentum as deeper and deeper layers of ice were ground into powder between the ship and the dock.

"By God, we got it," said one of the crew.

Another snatched up the bucket of whale tallow and slopped its congealed stench onto the rail ahead of the inching mass.

Josephine was standing well back, holding the child close and protecting her ears from the unearthly sounds. She had seen ropes snap under fatigue before and knew of the horror of their whiplash effect.

When the mass reached the end of the gangplank it was

already on the lip of the ship and the keel started to rise as the ship rolled dangerously sideways. The bowing crane jibs pulled from above to lift and steer the mass forward into the centre of the vessel, swinging it above the open hatchways, counterbalancing the capsizing effect. For what seemed a year, but was in fact a few brief seconds, it became a pendulum that ticked through all the known time of the universe while ignoring all the vectors of men. It was seeking equilibrium on the other side of the planet; a diving point where it might shed its gravity and exit to swim back into its natural element of the foreverness of night. Peary knew that it would never belong to this earth. It would wait for the world we know to disperse. To bubble and crumble away so that it might fall through all that we have known, all that we have been and will be to continue its journey elsewhere.

"LOWER," he bellowed.

The winch crunched into reverse and the great iron slowly made its way down into the hold. The sound was now a heavy breath of exhalation as the buoyant hollowness of the boat was replaced by the darkness from the stars.

"Turn it," the captain instructed the crew, who had ropes and staves of wood gingerly hooked to its descent.

Even the packing boulders of metamorphic rock cringed as the Tent swung above and was guided to nest implacably among their fracturing, helping bulk. Suddenly the steel chains and the iron cable slackened. The monster was at rest. The winch was shut off and a great silence slid over the dock. The iron was gone. It could no longer be seen from the land. The ship had swallowed it inside. Everyone waited. No one dared to speak, to breathe, they all listened. Expectation pressed against the white stillness, waiting to hear the keel snap, the ribs buckle and the *Hope* rend apart as the meteorite passed through its matchbox containment and continued its quest to the bottom of the frozen sea.

Instead it settled. He had it contained. Owned. The crew and the Inuits descended onto it, shifting the stones and the other packing to secure its agreed slumber.

PILGRIMS

McAlister's Godforsaken hutch was a long way from the gentle woodland and endless flat horizons of the Lowlands. 'God made the world, but the Dutch invented Holland.' The pride and the clarity of that statement exist inside the entire nation, and the truth of it is much stranger than its exaggeration. The reclaimed polders held back the salt tides by a steady and continual engineering that could only come from a patient mind. It also evolved in it determinism unlike anything else in the rest of Europe.

Van der Linden had truly become McAlister, not just in the adoption of the owner's name, but in the actual identity and rotting lineaments of the building. He rarely left it, and the scales that were once growing across his eyes receded, so that he could see all in the dimmed, varnished light of the shuttered interior. It had been two years since he had nailed them permanently shut and hung thick blankets across all the doors. The raging eye-peeling brightness of the outside was banished forever. It also gave him the advantage of being able to clearly see what manner of man or beast was gaining entrance through the swathes of woven barriers, while their scorched eyes were still adjusting to the dark. A few armed raiders' scalps had joined Tully's over the bar because their sight had been trapped between the gritty warp and the weft of the contrasting worlds while they forced entrance.

Skirmishes with dissatisfied customers and passing madmen were far too common in the parched midyear. Shotguns were all right when the rooms were mostly empty. But in a crowded swelter you needed something else to pick out the pests. So, in true Dutch pragmatism, the owner of McAlister's equipped himself against most forms of attack. He knew that in all armed forays a one-armed or weak-armed man was at his gravest disadvantage when it came to reloading. To this end he stashed three LeMat Grape-Shot pistols about the rooms to give detail to the shotgun's overall composition of blast. These massively constructed revolvers looked as if they had survived from the middle ages, and were grim and disapproving of any sleeker, more recently evolved design. The gun's unique feature was that it carried nine rounds in each of its cylinders. These did not rotate on an inert spindle, as did all other early revolvers, but spun around another active smoothbore barrel that could be loaded with all manner of shots and nails, giving the monster the double-jawed capability of a scattergun and a fairly accurate pistol. Three of these meant that a weak-armed man could muster a huge and continual firepower without reloading. The LeMat existed before the brass-cased bullet and it would take ten minutes to recharge and cap this triple-headed beast. So it was reassuring to know that at a pinch, when push came to shove and at close range, McAlister could always throw the five pound chunks of spent obstinate steel at his nearest enemy.

He had won the dragon-like weapons from one of the pilgrim guides called Dominic. He worked only once a year when his church ordered him to make the passage. He was hovering by, waiting for a sign of the arrival of the victims of the wagon train that had been allotted to him in order to ferry across the fire canyons.

"They's a different bunch this time. Your countrymen starting out afresh wid a some kinda new Gospil and a hope to guide 'em. They plan to grow corn in the wilderness."

He had McAlister's attention.

"I got word that they's supposed to dock up in Saint Simon's a month ago which makes 'em due around about now."

"Where's they from?"

"I got it writ here somewheres."

He poked about in his saddlebags that were piled up around his chair, in the corner of the room that he had won. Free board and lodging for a month or more was more valuable than a brace and a half of antique pistols. Eventually he pulled out a bundle of paper with a string around them.

"It's here somewheres."

He finally pulled out a scrap and passed it over. McAlister looked at it for a long time, then said,

"Krisitannia."

"That's the one."

"That ain't the Lowlands."

"No. It's in Dutch."

"No it ain't, it's Ultima Thule."

"Where?"

"The high North lands, mountains of snow and canyons of ice so I heard."

Dominic thought for a while and then huffed.

"Well it's all the same to me."

"This month, you say?"

"There or abouts."

"They're coming a long way."

"Aye, over deep seas."

Their conversation had been easy to that point, giving no reason for McAlister's wound to suddenly yank at him. But it did. He winced and turned pale at the sudden attack.

"You Ok?"

He could not speak, the air had been sucked out of him or maybe swallowed into the red hungry gulf in his body. He held up a hand and waved it in Dominic's face, trying to say that he was. Chava, who had been somewhere else, appeared in the room and quickly came to his side.

"He Ok?"

"No, what you do?"

"Nothing... We were just jawing about some folks we don't know."

Chava kicked things away to clear a space on the floor and lowered the ailing man down.

"He got some ague of the heart?"

Chava said nothing while he helped the crying man wind himself into a foetal ball.

"No more talk," he announced, clearly indicating that he believed it to be the instigator of the attack. Dominic was about to explain that he was innocent of any such malice, when he saw the look in the Indian's eyes and decided not to bother. He gathered his things and retreated to his corner, where in the dark he watched McAlister writhe like a beheaded snake until the Indian gave him some kind of draft or potion that put him out. After that the savage tenderly covered him completely with a thick patterned blanket.

Dominic's information about his expected customers was essentially flawed. The Norwegian pilgrims had fled starvation, bitter poverty and religious intolerance only two weeks before. They had crept from the sanctity of the safe natural harbour into the unforgiving North Sea, then into the deeper perils of the vast North Atlantic Ocean. Seasickness had sent eighty percent of them below deck in the first six hours, seeking something solid and unmoving in the massive swells. Many hugged the sealed barrels of Gudbrandsdal corn in the vain hope that its earthbound potency might soothe the rolling oceans into a sensible calmness. The precious kernels had been chosen for their steadfastness and nutritious potency, even though they had failed in the bleak years of Arctic misery. But in the new bright and promised land they would flourish and sprout with vigour and wealth. Had it not been prophesied so by the flock's pastor?

HELM

The white blueness sung into a transparency that cracked and spat like flames as the reinforced bow of the *Hope* splintered it asunder. The low icebergs no longer sent their tendrils and claws in search of Peary's bunk. Instead, they dug hard to sense where the iron rested. The interior voids and pits in the berg are always filled with noise. Pressure, sound and constant adjustments of trapped air being compressed by temperature and shifting gravity. But now they were hushed and listening, pressed against the ship's wooden chest to find the pulse of the brown solid heart. But the *Hope* would not tolerate that intimacy and shrugged aside the impudence in imperial rolling collision.

Peary was on deck watching the shadows of the departing mainland. The sea was rising to swallow the wind that had come out of the North, making the swell uneven and giving the small bergs a restless drunken motion. They steered by sight until they cleared the loose ice packs and lost all sense of land. Then they moved to the eye of the compass and its southern setting in the heart and volume of Baffin Bay.

"How is she fairing, Captain?" Peary shouted over the wind.

"Better than ever, the weight of that thing gives us a sincere footing and makes the old dame steady and sober."

"Excellent."

It was over twenty-four hours before the helm noticed that their astonishingly straight trajectory, and his almost effortless adjustments at the wheel to keep their compass bearing, was not luck, ability or the will of God, but error. The compass needle was held solid, strangely in the bearing they had to make. It had not moved since they left Greenland's shore. The Tent had snatched it for its own, its huge magnetism reaching up through the ship like an arrogant ghost. They were sailing blind into the Davis Straight and the Labrador Sea and the midnight forever expanse of the North Atlantic, with only the sleeping sun and the distant stars to guide them.

SIGHT

It was becoming clear that McAlister's malady was seriously affecting his mind, and that now he hardly left the slouching building at all. He had even developed an aversion to visit the outhouse and adapted a corner of the second room into his own personal privy. This added much to the interior's already pungent atmosphere. Chava thought it was the most disgusting thing that he had ever heard of, let alone witnessed. Even beyond all the other blasphemies that white men committed daily. His solution was to spend more time outside and only return to the innards when the weather hammered him or he needed a different kind of sleep. He changed his jobs for Mac; trapping daily and keeping the outside together and working; spending more time in the stables that were now cleaner than the rooms inside. On his way to and from the desert he tended the lazy sun baked horses, keeping them clean and watered. He preferred to sleep there when the weather permitted. The horse's shit smelt better than Mac's.

Dominic still waited for his pilgrims, playing cards with anybody who passed by and occasionally indulging in target practice with some of the arsenal that he kept in his saddlebags. He shot outback, where the dead were buried and the building absorbed some of the noise between him and the horses. The strange thing is that it was during one of these sessions that he thought he saw the first one arrive. McAlister was at the well. Chava was in the canyons and the only other paying customer, a man called Morton, had no interest in playing cards. He read books. And because McAlister would not open the shutters, and the daylight outside would reflect off the white of the page and sear his eyeballs, he had to do this by candlelight. Morton could also only understand his books if he read aloud, which he constantly did in a slow, lifeless murmur. This gave the fetid interior a solemn and church-like ambience that drove Dominic to distraction. So he selected one of his bags and hauled it outback to a low table in the shade of an ancient Juniper tree. He pulled out a canvas bag and unstrung it. Inside was a rifle in four pieces and a less than basic telescopic sight. It had once been a Martini-Henry cavalry carbine. One of its previous owners had decided to change the firearm from a perfectly serviceable weapon into a sawn-off clandestine tool of great clumsiness. Dominic hoped it had only been adapted this way by poachers, but deep down he recognised the

crude hallmarks of assassination. He quickly assembled the
blunt thing and took out one of its heavy cartridges that he
had bought with the thing. He worked the under lever and
loaded the brass into the chamber. Far out beyond the broken
fence was another Juniper tree. He squinted out of the shade
at its twisted dry trunk, then brought the weapon up to his
shoulder and looked down the dented metal tube. Its lenses
were misty and cracked, but it did bring the tree closer, so
when it stopped wavering with his breath he fired. It kicked
like a mule and it missed by a mile. He tried again three more
times without any apparent success. On the fourth attempt
he was just about to fire again when a blur filled the scope. A
blur that hovered in almost human features. He lowered the
gun and stared out to where it had been pointing. Nothing
was there. He moved out from beneath the tree and walked to
the fence. Nothing. No one was out there. The air was still as
dry as an autumn leaf. Nothing moved in its static thickness.
He very slowly brought the gun and the sight up to his eye,
keeping his other eye open for comparison. He scanned
the landscape and found the tree. Then saw the figure again.
It was a woman, a blur of a woman. She looked as if she was
flailing and soaking wet.

"I'll be damned."

He lowered the gun. The tree was alone again. He
detached the telescopic sight for fear of scaring the woman
with his gun. The sun was hurting his head and the tube was
slippery in his sweaty hands. Then he stabbed his sight out
undaunted to find her again. And he never did. He stood
staring until his eyes hurt. She did not exist and he would
never speak of her. He told himself it was an effect of the heat,
on him and on the land around, forcing up miasmas to fool
the vision of a simple man. He took the hot gun apart, slid

its component parts into the bag and retreated back into the sheltering darkness.

Two new men had arrived during his absence and were looking around the rooms. He could only see them as dense shadows. One was touching his saddlebags.

"You Yadill?" the other said.

"No I ain't, and get your hands out of my properties."

Dominic had a short-barrelled pistol holstered in his belt near his spine, and the wrapped disconnected carbine in his hands. He pulled it taut so that its barrel poked hard against the cloth, letting its identity be known. Even in the gloom both of the men understood its profile.

"It's all right friend, I was just a looking for a drink. Sees there ain't nobody here."

"Drink's behind the bar," Dominic said, his other hand already on his pistol.

"Where's Yadill? We's come to see him."

"Don't know no Yadill. It's only me, McAlister and his Injun dwelling here."

"That's the feller, they said that was his other name."

"Who said?"

The voice came from the back of the room. When it moved forward it was pointing one of the shotguns. Its face was green and the eyes were irregular. Large black smudges.

"OH GOD..." said the man near the saddlebags.

The figure came closer and Dominic hissed,

"Jesus Chava!"

The old Hopi ignored him and made the men disarm and say their names.

"Bone."

"Mulvanny."

"Chava, where's Mac?"

"Out."

"But he never goes out."

"Getting ready for the storm."

"What storm?"

"The one that's comin'. See my storm face."

"It's quiet and still out there..."

Chava laughed and gave him the shotgun while he picked up the intruders' weaponry.

THE STONE FENCE

Peary and the captain were staring at the unfolded maps. They were looking at the contours of the canyons in the Labrador Sea, thinking that if they could make out the western coast of Newfoundland then they might be able to hug it and follow it into the turn of the Eastern Shoals. That channel would take them home and avoid the icebergs that were sailing as if by a great will towards The Stone Fence; a long ridge of sub-oceanic distinction, seeking the depths of the Abyssal Plains under their massive whiteness. For days they watched and steered the *Hope* away from any of the albino mirages that threatened the horizons by day and night. Then, on the sixth day, out the fog came and the crew crossed themselves and stopped talking about the influence of the brooding shadow below deck. The captain posted men at the prow and they slowed the engines to a chugging crawl. The ballast that held the Tent occupied the areas nominally allocated for extra supplies of coal. The passage back had been calculated to the last ton. Being lost at sea was never seen as an option to such an experienced expedition. As the night congealed the fog, everybody prayed for a wind to cleanse the sea, but the heavy cloying obscurity just got thicker.

The ship that bore the Norwegians had braved the perilous depths and the mountainous waves. It had avoided the titan ice that sailed torpidly through the night, passing within forty kilometres of the very spot where, fifteen years later, an iceberg would slit open history and create the greatest shipwreck of all time.

They had survived and were approaching the shallowing of the Flemish Pass, where the remote sea floor rose up into sheer submarine cliff faces that squeezed and buckled towards land, the lip of the platform following the mouth of the abyss. The cracked contour of the drop and the rise were named into canyons: Carsons, Kettle, Hoyles, Denys, Whitborne, DesBarres. All the way to the encrusted epic ruff of the Laurentian Fan, its funnel shape looking and behaving like a gigantic drain sucking at the stability of the Labrador, and the broken coast of Nova Scotia. Near its edge at 43°57' 0" N 59°54' 57" W quivers one of the stranger oceanic phenomena in an ocean of perpetual strangeness. But it is invisible now, like all else in the closed pages of the dense fog. As the night waned deeper, the captain, who had been steering by his resolute compass, became horrified to see the needle deflect and shudder as if it were being flicked or tugged mischievously away from its normal stoic duty. He called his mate to witness it. They were within three-hundred kilometres of the mainland, less than half a day to salvation. But there was no sense of that among the pilgrims, and even the sea-hardened crew couldn't get a whiff of terra firma through the suffocating air. An hour later the captain realised that the compass could no longer be trusted; its erratic lurches were jesting with the maps. That's when he began to pray.

SAND

The red wind hit McAlister's with a vindication that rattled the dry roots of the hutch and nearly snapped off all the bristles that burrowed and anchored it to the sun-fired clay. Even the water in the well, that had never seen the light of day or the dark of the night, trembled. Chava feared for the horses. He had lowered the rush screens and closed the stable doors, but the wind could be heard moaning as it fumbled with the clattering latches. The sand had been carried high and filtered the light of the sun against its red lens, so that the normal spectrum was warped and slid eerily towards sickly greens and slate greys. The sky above the wind was orange and mercifully out of sight.

The men inside were speaking to each other again, and they even tolerated the old Indian's commands about what to do to reinforce the shutters. Everything was battened down except for the main door. The sand was pushing in under it like parched lava while they waited for the owner to return.

Mac appeared just before the sound of the door being scratched apart forced the men to barricade it. He was erratic and half dressed, his face masked in a wet bandana and his Queen Victoria goggles. He was walking crab-like, because the pain in his shoulder made him stoop like an old man. His clothes were encrusted with sand but his head was wet and dripping onto his arid shirt, making dark coin-like patches which matched the metal spectacles he wore over his abased face. He was barely in the door when Bone spoke loudly over the wind and the noise of Chava sealing the door.

"You McAlister, called Yadill?"

"None of them," he answered without looking at the questioner or anybody else. "None of them." He said this in a voice that made Chava leave the door and hide the stranger's guns deeper in his chosen seclusion. He then slowly sat and watched 'his master' very closely, especially as he was making his stumbling way behind the bar, where his own arsenal was stashed. Dominic was in his dark corner with one of the saddlebags on his lap, his hand inside it.

Mac ducked down behind the bar and Chava held his breath, convinced he was either ready for gunplay or about to perform his ritual snake dance again. Then he reappeared like a jack-in-the-box with a bottle of whiskey in each hand. The bandana and the spectacles were gone and he wore a wide and previously unseen grin.

"Blowin' up out there. We be here for days. We had better get stuck in."

The men looked at each other and then agreed. The idea of whiskey and sanctuary was greater than anything else they could recall. Mulvanny was first at the bar.

BLIND

The *Hope* had dissolved. Not in an impressionistic haze of shadow and particle, but more like a haunch of flesh and bone in unrelenting acid. The night fog had reached its darkest, a condition that was the exact opposite of invisibility. There was no point in the lookouts straining their eyes at the prow of the ship. There was no space before them or behind. Only the motion of the sea and the throb of the engine convinced all aboard that dimension did exist somewhere, even if they were not a part of it. A thick, salty blindness separated all. So that nobody could see their companion. Even the whale oil lamps of masters' cabins had become smothered and seemed to want to return to their hidden blubber origins. The salt of the ocean had solidified around them and the only thing that now bound them was a great thirst. It threaded them together like shabby pearls on a withered thread that wanted to break over the idea of waves.

Everything inside McAlister's trembled. The men sat together around a circular table passing the bottles back and forth. Bone could not leave the other business alone and worried at it, even though he and everybody else knew it would lead to blood. The room and everything else was already tinted red. Outside you could not see your hand in front of your face, so thick was the whistle-sagging air. The horses had stopped kicking and whinnying and now they were silent and morose, breathing in the saturated wind. Everything rattled and moved and yet the overpowering atmosphere was of suffocating stillness. Chava knew it was best to sleep or hide until it had blown over. Talking through this was dangerous. That is why he stayed, because he saw that McAlister was transforming, becoming reborn again. The Hopi had a complex belief system that told of continual generation into new world through a system of enforced hibernation. They had dug deep into the land, like ants, and would seal themselves in kivas; cysts of time under the world. When the time was right, when the omens were bright and their hibernation had transformed them, they would break the seal and return again. Waking in a new world. This metamorphic reincarnation had been the machine of their evolution since they were created. Chava had never seen the slumped buildings of the trading post as anything like a kiva, but now he had his doubts. White men do everything wrong, sometimes backwards; this would explain everything about the Dutchman.

"So you ain't Yadill then?" Bone said.

"Or anythink like it?" chirped in Mulvanny.

Dominic groaned and Chava made ready. McAlister though for a while, rubbing his chin in stagey consideration.

"Well... I ain't. But my wound might be."

Then he pointed at Chava.

"He says it might be the name of my wound's father."

Mulvanny looked at Bone, searching for any sign of clarity on his face. The red wind shouldered the door and the shuttered windows and moaned eerily down the chimney. It had been built with iron running across its aperture. So fierce were the attacks in those early days that access had to be barred in all quarters. Bone continued, choosing to ignore the hogwash that had just been spoken.

"You ain't no Sven Yaddol either?"

"Cause that's the man we seek," butted in Mulvanny.

"E's a foreigner like you and can't speak American in a good way."

"That's a Swedish name," Dominic said. "Mac here comes from the Lowlands of the Hollands."

"Who asked you?" said Mulvanny, while Bone stared aggressively across the table. "Dutch, Swedes, all the same to me. Are you him?"

"Aldrig hörde av honom," said Mac grinning.

"What he say?" said Mulvanny. Nobody knew. "Sven Yaddol killed our kin. We bin seeking him ever since. Some trader told us about a man here bearing the name."

"Shut up Mulvanny," said Bone.

"But it's him, ain't it?"

"I don't know, but he will do."

Bone was out of his seat, drawing a concealed knife. Mac pointed at Bone, but addressed Chava.

"Another wig for the wall," he said.

The *Hope* was closer to the shore and running parallel to the Stone Fence. But nobody knew that. Peary sat with his family, and now held hands since the yellow lamps faded around them. They waited in the congealed tension for the sound of collision or the shock of the jealous land to claw through the ship's thin skin and snatch the great weight free. Josephine knew how dangerous things were because her husband's lisp had returned permanently. And even though he had tried to hide it, or to not speak at all, it was apparent. When she asked him direct questions, he mumbled or excused himself. Now he just sat morosely, pulling and twisting the ends of his long moustache.

At eight bells, a late dawn was warming the air outside and the first optimistic signs of a clearing could be smelt in the rigging. That's when the sound came to the *Hope*: a far-off blunt screaming that sounded like an axe was being sharpened on an erroneous, complaining stone. Or a tree being sawn open with a fresh tusk. Its wrongness was worse than its pitch and timbre, which made the skin crawl. No good could ever come from hearing such a thing, and those that did covered their ears with their hands without knowing why.

Outside, the storm finally worked the latch and the stable doors flew open. The angry wind immediately lashed a great suction inside the hutch and blew out all the other shutters. The horses went mad, kicking and bucking against their tethers and restraining stalls. One escaped and forced its way outside, only to be blown sideways against the building. It had been tethered on a long grazing lead that Chava had not bothered to take off when the storm arrived. The beast was outside the main door of McAlister's when the lead reached its length and tugged the horse to a halt. It tried to pull itself loose but only managed to hook the thrashing leather around some solid feature of the building. The wind was driving the sand against the flinching animal, making it blind and demented, and forcing its violent movements to twist the lead and strangle its freedom of movement. It was crashing against the door and outside wall as the wind crushed it and the tether pinned it there. Some of the hair of its mane and tail had already been shredded away. The men were mesmerised by the sound of its terror. Then something like a splash hit the far side of the table, an abnormality of cold in the burning air. Dominic and Bone jumped back, as if hit by a wave.

"Fuck that!" said Dominic and Bone almost dropped the blade.

Mac laughed in between the pain of his roaring shoulder, until he saw the first one standing across the room behind them. He then reached under the table and pulled one of the

LeMats from its sprung clips there. Another, a woman, now stood behind the first. They were lost and carried it like a virulent contagion, their terrible eyes peering hard and only just seeing Mac, as if he were a century away. Mac lifted the massive pistol, cocked its serpentine hammer and fired across the room. The ball went between Dominic and Bone and blew a lump out of the wall.

"Fuck," said Dominic, and he threw himself to the floor.

Bone charged at Mac as he re-cocked and got the next round in his enemy's mouth. It split his palette and creamed his cerebellum, exiting below his left ear. Mac kept firing at nothing until the gun was empty. He then ran back behind the bar and retrieved a shotgun. The whirlwind of glass-sharp sand had abraded away the first layer of skin from the exposed side of the horse and was gritting its way into the red raw open muscle. The horse kicked, stumbled and threw itself against the building in agony as its anatomy was being peeled away. A living Stubbs painting, sandwiched and crashing insanely between the storm and the imprisoned men.

The impossible phenomena that lived just beyond the Stone Fence was called Sable Island, a shape-shifting sandbank. Twenty-five miles long and one mile wide. It is most often described as an irregular sliver of crescent moon, or a lost pairing of a giant's fingernail. But its contours vary and change, and it has even shifted position over the centuries. Its frail existence should have been brushed aside or eroded by the tempestuous seas that surround it. But it lives on as a lethal geological enigma, shipwrecking hundreds of vessels. The only protruding obstacle in an otherwise open sea, earning it and the surrounding waters the nickname of 'the graveyard of the Atlantic'.

The Norwegian ship did not have a chance, its spine was snapped and every soul on board drowned in minutes.

The blast from the shotgun blew a hole in the wall the size of a dinner plate. The sandstorm found it instantly and sent in a dervish spiralled beam of light, wind and sand. It passed straight through the apparitions, who had multiplied, so that Mac could see over ten milling in and out of the swirling spotlight. Some had been smiling as if they were pleased to be there. Then after a while, as their eyes cleared, they saw where they really were and the thin candle of hope was blown out forever. The horse had stopped moving outside. It was still alive but blood loss and shock had condemned it to lie against the wall before exhaustion made it fall on to the exposed bones of its knees and fetlocks. Dominic and Mulvanny where still crawling about under the table, convinced that Mac had turned insane and was after their lives. Chava stood far back in the room and watched everything as if from a great distance. There was some kind of meaning in all of this and he was waiting to receive its wisdom.

The dense fog cleared and the *Hope* made its way by straining eye along the exaggerated coast until it reached Nova Scotia, where cross winds cleared the last remnants of hazy blindness and gave the ship's master permanent sight of the land and stars. Everybody breathed again and came up on deck to sense the depth of the flowing days and the crystal eternity of night.

The apparitions discovered that they could not put down their tools and that the yellow of the cornfields of their dreams, in which they had awoke, was in fact the yellowness of the sand that sucked and polished their bones. One or two of them could see McAlister, but only as a far away blur. His shoulder was agonising and it had acquired a dense top-heavy weight that made him totter every time he discharged a round, which he finally realised was pointless.

He turned his imploring eyes to Chava. The room was filled with rods of spinning light that the bullet holes in the walls had given access to. And to a whistling that came from their ragged lips. It was a high, white sound set against the low, purple moaning of everything else. Outside the horse was totally flayed, dead and adhered to the wooden slatted wall. Suddenly and without any kind of warning the wind reversed, halted into something that might be considered its opposite: a solid stillness. The heat glared on, but the atmosphere was motionless. A tiny trickle of sand out of the eye sockets of the horse was the only movement. Inside, the furnishings and objects of human occupation slumped into stasis, exhausted from the dance of shuddering and flinching. The only sound was the breathing of the living men, which was slow in catching up with the quietness of everything else. The ghosts must have faded in the parallel, giving up their tenure and image with the softening of Mac's breath.

"Chava, is it safe?" Dominic asked in a whisper from beneath the table.

The Hopi looked into McAlister's eyes and after a while said,

"Yes, you come out now."

Nobody moved.

"Its Ok, they's gone," said McAlister. "I ain't gonna shoot no more if you don't."

Mulvanny and Dominic started moving out of sight, shuffling across the floor. Mac held his arms out straight so that all could see he was unarmed. The only trouble was that now he looked like a sleepwalker, which was almost as bad.

"You better get whiskey," said Chava.

Mac agreed and turned back into the bar to retrieve a bottle and glasses. Dominic arose behind Chava, the small gleaming pistol still in his hand. Chava glanced back.

"Put it away, no more killin' today."

Mulvanny was kneeling by his dead comrade, looking at the mess that the crazy Swede had made of Bone's head.

"No more killin'," Chava said again, and Mac opened the bottle.

BALLAST

There was a human ballast on the *Hope*, just in case the Ahnighito was not enough to establish Peary's scientific pedigree. Five barrels of Inuit bones had been 'excavated' from their carefully composed sleep. And four and two 'half' specimens of his 'Little Brown Children' were being brought back for examination. Precursors, living prototypes of the horse hair and wax versions of them that were planned to grace the American Museum of Natural History's Arctic galleries. Franz Boaz - the 'Father of American Anthropology' - had been given an exalted position and exuberant funds to flesh out the new Museum. He had asked Peary to bring one living Inuit for him to study, not two and a half families. They had been persuaded that it would be a good and important thing to do; to go to the land where dwellings touched the clouds, sledges ran on wheels without dogs and light was owned and controlled by the thousands of people who lived and hunted there. They had not been told about the barrels of their ancestors.

Qisuk and his seven year old son Minik; Nuktaq, his wife
Atangana and their infant Aviaq sat huddled with the lone
ranger, Uisaakassak, against the fog and tried not to think
about the stolen demon which occupied the core of the ship.
They had helped Peary find it and were there every year of
his returns. They had no fear of waves and ice, but distrusted
largeness and duration. The *Hope* and its forever journey
weighted heavily on their living instrumentation of the lands
they were leaving. When they felt the Laurentian Fan beneath
them, emptiness deeper than all those waters flooded up to
extinguish the lamp of knowing in their stoic ribcages. When
the fog dispersed and the edge of naked land was spied they
shivered at its strangeness. The relief in all the white and black
men compounded their anxieties. The tangible breath and the
living bone of the great spirits were draining away, and the hard
and meaningless rocks that replaced them were without soul or
significance. They spoke of it being the price that Tornaarsuk
the evil one demanded. But not all agreed. One and its half
were beginning to think that Ahnighito might be the only part
of them that they would know in the new place to which they
were being taken. Had not their fathers and all their fathers
before them chipped blades off it, and its Woman and its Dog?
There was reality in those acts and reality could always be
trusted, maybe even in the land they were being taken to.

When they eventually moored in Brooklyn's Naval Dockyard,
they knew that they had been wrong.

GIN
Mulvanny left McAlister's without ever looking at or into
the proprietor's eyes. Nor did he acknowledge the dead

horse which was smeared into the outside wall, or the re-written landscape that he was about to join. He took up some of Bone's possessions and their remaining sane horse and disappeared into the direction where he thought the track had been before. He would always tell those rich and thirsty enough to listen about his showdown with Mad Sven Yaddol, and how they had had a raging gunfight in a sandstorm, and how he had slain the butcher with a single shot through the mouth and that nobody would ever find the body because the wind and sand had ground it to dust so that it could be blown away by a malicious and justified wind. Dominic left a day later, vowing never to return until McAlister's had revolved again and the next one was of a more stable temperament. Chava buried Bone out back amid the waste of splintered crucifixes, and did not bother to add another.

That night was very silent, there was no wind and the coyotes too enjoyed the peaceful stars after the raging storm. Mac and Chava sat at the same table eating a broth of beans and wild turkey. Mac had opened a crate that Chava had never seen before. He pulled from it a long earthen flask, also of a kind that the Hopi had never seen. A coloured label with words was pasted on. Mac uncorked it and poured two glasses of the colourless liquid.

"Tequli?" Chava said.

"Nee meneer het is pure jonge klare."

Chava smelt the contents of the glass, then cautiously sipped at it, a smile broadening on his face as he did so.

"I am never going outside again," Mac vowed.

The old Indian nodded as if he understood. They drank two stone bottles of the fine spirit and fell asleep across the table, as the silent night slid out and a soft muffled day sidled in.

DOWRY

Boaz received his specimens days after they nervously landed. He had forgotten that he had asked Peary to retrieve one for him. It had been two years since he had mentioned it, and his work at the Museum had been filled with organising expeditions and retrievals from all over the world. As he stared at the wide-eyed band of Greenlanders, his heart grew faint. They were already in the public eye and he no idea what to do with them.

The iron Tent sat in the Brooklyn dockyards for seven years, the grey concrete beneath it turning ochre red while Peary, and later his wife, argued about its value and its price with the Museum authorities. The Great Iron Tent and the barrels of bones were to pay for his next expedition. Stepping-stones to the Pole and time were running out. The obstinacy of the Museum to meet his price sometimes seemed to match that of the Ahnighito. Peary liked to barter, but this was getting nowhere so he lost interest with the white (brown) elephant and gave it to his wife as a gift, partially as a guilt dowry on his other spouse.

Josephine had found out about Alakahsingwah. She had personally met her on one of her own unannounced trips to visit his Greenland camp. She had met her child and seen the cast of its eyes. Her fury was almost as powerful as his will, but had nothing of its relentless velocity. He patiently explained to her, like he had to others, that the Eskimo women should not be considered like others. They had none of the natural modesty and moral fortitude that all civilised women were blessed with from birth. Instead they had a comely animal vigour that was natural in barbaric and brutal lands. They had physical instinct about the needs of men that was invaluable to explorers like he and his men. Peary had often exchanged material and trinkets with his sailors for the use of the village women of the Inuit encampments that had so trusted him. There is even a photograph, taken on one of his ships. It is different from the anthropological studies and his poised portraits that were made at the same time. It shows six local women dancing in a high step chorus line, their wide smiles looking as artificial and frozen as the can-can they have been told to perform on the icy deck. Josephine bit her lip as he explained that he had no intention of giving up an important resource to his health and well being when travelling the Arctic wilderness. The matter was closed. Better to put all thoughts about it behind them. He did not want unhappiness and quarrelsome bleakness to sully their contented home at the Eagle's Nest.

Josephine never visited her beloved Greenland again. She finally sold it to the Museum for a meagre $40,000.

During the time of negotiation all of Boaz's specimens had died. Except Qisik's son. Minor ailments, homesickness and depression snuffed out the brightness of their first glee at seeing New York. Minik had watched his father fade to nothing in hospital, and feared for his soul without a shaman. Boaz took the body and promised the boy a respectful ceremony, and he kept his word in the form of a play. He arranged the gathering around a newly dug hole in the impressive Museum grounds. The child held the hands of the Wallace's, who had taken him in. He watched Boaz carefully as all the words were said and thought about his beloved father sleeping under this strange green place, instead of the natural white of his home. He hoped that the words, although wrong, would allow his spirit to travel back there while his bones rested here. It could be possible if the balance was right.

From the third floor window of the grand museum the little party below looked meaningless and lost on the lawn. From the shelves at the back of the same room the window looked glaringly bright, even though it was permanently blinded. From the labelled boxes on the shelves the outside world was nothing but an indistinct muffled sound. From Qisik's bones inside the box, nothing could be sensed or felt. The acid that, a few days before, had stripped all the flesh had also removed any trace of memory. Scrubbed clean, the marrow dead and labelled, he had been secretly folded into nothing for future research.

Minik was blighted and desperate to return home. Boaz had even contacted Peary for help, but his interest lay further North. He was determined to scratch his initials on the North Pole, like he had on the meteorite when he first found it.

PETRIFIED

Before the hour when the sun showed its claws, Chava was outside and, amazingly, without a crippling hangover. Peeling and shovelling the horse off the wall with a pounding head and a rising gorge would not have been possible, even for a such a world-beaten soul as he. Mac still slept, the gin and painless shoulder giving him a space to drift deeply into for a few hours. Chava had laid him out on the floor, a flour sack under his head and an old clan blanket thrown across him.

Half way through his task with the horse the old man stopped and moved away, wanting to breathe clean air for a while. He walked a few paces away and looked at the warming land around. Then he turned back to the slumped building and was surprised by what he saw. He had cleared the bulky mess of the horse away and the spine had snapped in the process, so that now only a third of it was visible sticking out of the front section, which he had not yet touched. The front legs, a small part of flattened chest and the head were still there, hammered and glued into the tar-covered woodwork by sand, blood and pressure. It looked like some kind of exotic heraldic beast or one of the ancient creatures that sometimes they found trapped in the stones of the desert. A race before men, petrified forever for some unspeakable sin against their Gods. Chava's grandfather had found one of enormous size, with jaws big enough to devour a man whole. This two-legged fragment of horse with its smeared, dislocated skull and its spine like a tail seemed to be guarding the door of McAlister's; a fearsome demon out of the sandstorm had come to abide with them. Chava went back and cleared the stinking mass away so that he could examine it more closely. There was only one layer of it that was still moist and might cause a disconnection in its grip to the wood. But he knew how to fix that. He went into the stables and brought a half-gallon pail of creosote and a stiff brush. He mixed in some hoof varnish that they used to strengthen the horses' feet throughout the furnace seasons. He painted the sticky pungent goo all over the remnant of the corpse and worked it in hard around the still soft tissue. The sun was scalding him and his paint was drying hard on the brush as he used it, so he splashed and threw the last dregs over the carcass so that they ran and sizzled into long spiky stains. He rubbed his stained hands on his canvas work trousers and stood back to view his masterpiece. He saw that it was a fine thing.

Like all great artists, he felt the need to sign, not with his name, but its own; the name that nobody understood or wanted. So, before it completely dried, he took up a big bent nail and scrawled into its neck: YADILL.

PROCESSION

Thirty horses were harnessed to the iron carriage with solid steel wheels. The Tent was lifted onto it by one of the dock's tall cranes. It settled on its massive chariot without a whisper. The hostlers geed up the beasts, and the long procession creaked and strained. Wedges and levers were brought to encourage the wheels to move, and inch by inch the great weight of the heavens rolled into motion up 50th Street. A crowd followed the ponderous journey, which moved at the same speed as walking. The turn into 8th Avenue was tricky, and took all the skill and experience of the truckmen to achieve it smoothly. Some of the crowd whistled as the hardworking horses strained in unison. By the time it reached Central Park West a momentum was building and the brakeman had his huge hands fixed and ready on the ratchet lever.

Minik had to trot to keep up; his small bowed legs were not grown for this rigid terrain. He had followed it all the way, never losing sight of the Ahnighito, even when he was pressed out of the way and obscured by bigger people. All the people here were growing taller to fit inside the buildings that were also growing taller. He knew that one day the buildings would only hold four or five of them the same size; a race of long, thin giants standing upright in long, thin buildings that would remake the canyons of brick, steel and glass through which he now walked and whose wonder he no longer grinned at. Now his eyes were mainly set on the pavement.

When they reached the Museum and the unloading ramp, the crowd was told that the Tent would not be unloaded that day. The hostlers unhitched their beasts and disconnected the tethers and shafts. People started to drift away as the drama was taken apart, and only the mass of iron waited at the gates of its enthronement. The foundations of the hall that would contain the treasure had been reinforced to hold the Tent for a long time. Minik made his way towards the Ahnighito, the one physical part of his homeland that he could trust, the only thing for thousands of miles that shared something of his scared heart. They had been stolen together. Shared the darkness of the inner ship, the stormy water and bleakness of this land that had only been made by men.

He was within feet of touching it when a hoard rushed past him, pushing him aside; wild young men, feral children from the Bowery, barbarians nurtured on the pap of poverty and crime. A thin grey milk full of grit and immigrant envy. They swarmed over it, taking out jack knives and stilettos, trying to hack parts of it off for themselves. The blades of the cheap knives snapped, some taking the blood of their owners with them. Minik was amazed that they were acting like his people. But why? They already had metal blades and guns. And they hunted nothing. The Museum officials screamed at the thugs, calling the police to see them off, which they did with truncheons and metal handcuffs clipped over their leather gloves. The whelping youths fled and fell from the Tent under the vigorous attack. One of the boys hurtled into Minik and knocked him to the ground. The boy had red hair and piggish features that Minik knew were called 'Ironish': a tribe from another country that some said was colder and emptier than his own. In his time on the streets of New York he had learnt to stay clear of all those from 'Ironland', for they had a violent dislike of him and all those like him, whom they had never met. The boy vanished and a huge policeman now stood over Minik. He too had the same features. The look of the 'Ironish'. Minik covered his face and crawled into a tight ball.

"Leave him," shouted Boaz, who had been standing with the Museum officials.

The policeman spat and lowered his blunt harpoon.

POLDER

Mac was as good as his word; he never went outside again, and over the next year Chava even got used to him using the dirt box every day. The usual assortment of travellers came and went. Some brought news, some brought trouble. Fewer stayed forever. Chava's backyard spade was becoming as broken and as lazy as the two men. The wagon trains of those seeking the other side of the canyons came less and less. The times in the world beyond were changing. The great need to own new lands had become tired and the railroad had made legal boundaries across the wildest distances.

On a morning in a month and a year that they never knew, Mac called the old Indian over to the bar.

"Feel this."

He undid his shirtfront, baring the scar tissue for the first time since his operation.

"Touch it, it won't bite."

Chava put his gnarled and horny hand on the wound.

"Feel that, it's got solid again."

Mac's eyes were different; softness was there when he spoke of the change in his wound.

"It's like it's come back. The part of me that went missing. I ain't got no hollow anymore."

Chava said nothing.

"I think it's time for me to go home."

Chava waited.

"I'm gonna build me a polder, right here."

"Polder?"

"Yup! I guess it will be like them kivas you told me of."

Chava thought for a while and said nothing.

"Will you help me make a polder, build a dijk in the time of the next storm?"

They were still outside the storm season and Chava guessed that the first wet winds were at least a month away.

"Well, will ya?"

"For Jenever I will."

Both men laughed.

FEW

Peary had returned and was elsewhere in the city, trying to raise funds from the continually uninterested sponsors for his next expedition. His first attempt on the Pole had failed, but he had got closer to it than ever before. He had smelt it, the scent of its presence was now locked in his head. Determination had frozen into obsession. Valour into fascination. The Arctic had beaten him back again, but this time it was different; it was not the grandeur of Greenland, its kind terra firma, but the featureless harshness of the frozen ocean that would have nothing to do with men. It had turned his feet into numb wood, so that when his boots were unthawed and pulled off, three toes snapped off with them and remained adhered to the icy leather. Others had to be pruned off later before they blackened into gangrene.

"A few toes aren't much to give to achieve the Pole" he said in the stoic understatement that would inspire American youth for decades and ensure that his statue was raised.

One of his appointments that day had been delayed by the caravanserai on 8th Street. He limped to the window with the aid of a silver-topped cane. The tall building he was in was on the other side of the park. He looked out towards the Museum, somewhere hidden in the trees. The sound of the traffic had faded under growing leaden skies and the world seemed to be becoming pensive and still.

After the police and crowds moved away and Boaz had joined the other gentry and returned to the Museum, Minik approached the Ahnighito and felt its reassuring coldness. He closed his eyes and thought of the white mountains, the blue glacier and all the dogs. He said his father's name as snow began to fall around him and people hurried through Central Park to be home before the weather got any 'worse'. As it touched his face he said the names of those others who had been stolen with him and brought here to die. He bit his mouth thinking about the barrels of bones of all those who had been divided from their sleeping souls, and was grateful that at least his father was treated with respect by these cruel ones. The snow was now falling faster and he began to collect shards of the broken jackknife blade that the Bowery Boys had left under the police rout. They were a sign; two metals balancing the meaning of the journey of the Ahnighito. He would take them back and give them to the children of his people, so that they might use them to tip their harpoons and tools. He would no longer go to the school where the children pinched his body and adults pinched his mind. He was going home.

BY DAWN

They both felt the storm wetly smouldering away in the North three days before it hit. Mac told his one remaining customer that he had to go before it arrived.

"Go where?" the man asked.

"Anywheres but here."

Eventually he had to see him off with one of the massive LeMats sticking out of his scrawny belt. The oppression of the skies and the parched lust of the desert formed an in between space in which no men should be arguing. So the man left and headed Northwest to skirt the blackening sky and the pulsing yellow wilderness.

Mac gave instructions about opening all the doors and windows without ever setting foot outside. Dry powdery sand slid from the shutters as they creaked open, and the light hesitated before it flooded in.

"We gotta make lots of bundles of rags, more than ever before, and start diggin' the ditches."

"We?" said Chava, but Mac ignored it.

The ditches and bundles were collected to steer the gallons of water that would fall out of the sky and guide it into the well. While working with the heap of old dry sacks Chava decided that he had to make a protection for the protector of the house. So he took up some of the newest sacks that still had the fine smell of hemp radiating from their woven roughness and nailed them over the Yadill, covering its head, body and legs. The other sacks were rolled and placed into the patterns which Mac had described, then they were held down by large stones that would anchor them against the expected oncoming tides. These causeways lead through the rooms into the back, where the mouth of the well waited.

All the 'furnishings' that were capable of being moved, were. Some stayed for their season wash. All the dry goods were piled high on tables and improvised stretchers. Mac emptied his dirt box; it was the closest he got to the outside during the entire process. Chava wanted nothing to do with such a disgusting act, but agreed to tow the waste away into the desert if - and only if - Mac sacked it up and tied it closed.

After the long day's work they opened one of the last bottles of gin. The night flickered continuously, sending increasingly brighter splutters of light to ricochet around the irregular, congested rooms. The air cooled and whispered at the edges of the fast illuminations, playing in the eddies and mimicking the quick surprise. Chava had climbed up onto the sagging roof to watch the storm arrive. The lightning was almost constant on the dark horizon that bristled into purples, blues and whites with each fork stab. By dawn, he thought. The thunder will be here by dawn.

The blue-black voice mumbled above the rising sun, then hit with a volume that spilt Mac out of his bunk. The second and third shook McAlister's with a force that de-nailed the wood and kicked every loose thing asunder. Chava had given the horses a mixture of peyote and soothing brush so that the violent rages of the sky passed them by as they flinched and shuddered in time with a calmer day in a dreamt meadow.

The wind seemed to be coming from the land, not the slate steel clouds. It harried and swirled the dust and sand in defiance of the drowning rain. Chava had tied a bandana around his face, covering his mouth and nose. His ancient flapping Stetson was tied down hard on his head, which he kept bowed as he ran bow-legged back to the rooms.

The rain hit all at once, there being only seconds between the joy of the first heavy drops and the sky opening in an overpowering deluge. Chava had just made the doorway before the impact of water was illuminated by jagged, blinding lightning and applauded by a vast roll of deafening thunder immediately overhead. He did not know what to cover first, his eyes or his ears, so he just pulled his hat down further and rushed into the sweating cave that was giving up all its human stink in one long exhalation. Mac was standing behind the bar with a shovel in his hands.

The air cooled almost to chill as thousands of gallons of water fell and churned up the sand and were swallowed by the insatiable land. On the other side of the solid screen of rain, the entire landscape had changed colour. Deep blood-reds and blacks had knotted where before there were only searing oranges and sunken yellows. Every arid pore sucked at the water and gave up its constancy as payment. The glistening ranges knew nothing, nothing of parched deserts, brimstone, clinker valleys and furnace roads. Everything dreamed in cool green as the water flowed and hissed. The iron hard rocks cracked and opened their innards to own more rain.

Mac and Chava's irrigation channels and dykes were working, the bubbling streams were being diverted into a lake-like polder inside the rooms. The men were furiously wrangling and persuading the now calf-deep water to find its way a few degrees lower into the open mouth of the well. At first it was reluctant and wanted to chew at anything absorbent on its way. Then a trickle being swept on by Mac's shovel and Chava's broom found the incline and dribbled into the deep throat. The sound of it chuckling below was heard by the rest of the stream, which then rushed into the echoing depth. A sonorous and satisfying resonance came up, and its rich tremulous sound made the men stop work and their hearts heal.

It rained for four days, gradually slowing on the third. They slept in hammocks tucked up in the lopsided ceiling of the cleansed house. Their sleep was profound and refreshing, and Mac never woke up.

Chava buried him out back on the sixth day, giving

enough time for all of the soul to depart, to drain off
elsewhere. The sky was clear and the heat rose up moist and
scented with the earth. Amazingly, the ground was already
turning back to sand. The enormous volumes of water were
being stored in subterranean cisterns or gullied away into far
off rivers. The immediate canyons were drying. Chava dug
deep before the heat made the earth rock hard again. The
spot he found was to the side of the tide of broken stumps,
and away from the unmarked enemies, and in the shade of
the Juniper. He put the old Dutchman into the hole wrapped
up in their best native blanket, the LeMats placed at his side.
He slid the earth and sand back into the hole and onto his
friend, and then patted the ground even and flat with the
back of the spade.

Chava went back into the gently steaming rooms and
climbed high up into a dark corner where he kept his own
tiny stash of treasures. He brought down an old metal canister
with a tight-fitting lid, prized it open and took out a rumpled
fistful of something. He gathered a clean splint of wood and
tied the crusty ball of stiff paper to it with a short length of
bailing wire. When he was sure it was secure he planted it into
the grave, said something under his thick breath and went
back into the house.

He waited for the next McAlister to arrive for over two
weeks, then decided it was time for him to make a return. He
gathered his few possessions, chose the best horse and rode out
of McAlister's towards Oraibi, without ever looking back.

The last remaining dampness clung well to the Juniper tree,
whose roots had drunk long and hard in the last week. There
was even a stirring among the leaves as if buds or new shoots
were considering coming out to see what kind of place

they might dwell in. The damp also affected the wired ball
of stained paper, making some of it unfold. Making some
of the dried blood fall away. The wind did the rest, teasing
at its confinement, worrying at its unity. It was opening out
in a weird blooming towards dispersal. Just before it was
finally shredded and blown away into the forgetfulness of
all things, its inner folds came apart so that words might
be seen there, barely clinging to the frayed surface. Some
of the words told of new inventions in the East, some of
great stones from the sky being found in the high North.
One set of words told of the old West. There had been
more dried blood on those words, but some of the bolder
headlines could still been seen. The largest told of the death of
HENRY (DOC) HO... the legendary friend of Wyatt Earp,
and of his place in the history of the West. It was difficult to
read because part had been erased. Even part of Doc's name
was gone.

END

Robert Edwin Peary finally conquered the geographic North
Pole on April 6th 1909. He had reached his goal with no toes,
having the few healthy remaining ones removed so that
he could learn to walk without them. He had achieved his
ambition and the recognition he so desired. He retired to
Eagle's Nest with Josephine, where they lived until his early
death at the age of sixty-four. The Arctic had taken its toll. He
was prematurely aged and divorced from the great whiteness
that illuminated his existence.

There was some disquiet about his crowning achievement
in his later years, but it is fortunate that he died before all the
serious controversy arose about the validity of his claim to

be the first man at the Pole. There can be no doubt about his fortitude and ruthless determination, but his ultimate success is widely doubted today.

As Peary was supposedly planting Old Glory in the ice, Minik made plans to get back home. He had found out the truth about his father's bones and it drove him under. Eventually he wrote a letter that brimmed with disgust and his sense of betrayal and gave all the details of how he and his people had been treated.

The *San Francisco Examiner* took up his case. The public was now on his side and Josephine Peary kindly raised funds to send him home, but only after signing a paper saying that he would never return. In Melville bay he met Peary again, who was returning from his 'triumph' at the Pole. Peary made Minik sign another paper before he was allowed to continue

further, which stated what he had done: to help the young
man and exonerate himself from all future blemish and blame.

It never really worked back home. Something of Minik's
contentment had been erased, and a lost longing replaced it.
He was permanently imprisoned in the fulcrum of everything
he had been forced to see and survive in. In 1917 he returned
to the US, wanting to talk more about his plight. But the
Great War was raging in Europe and all eyes were focused in
that direction. He worked as a lumberjack until the great flu
epidemic swept the county. He died in October of 1918.

A few years ago, the Museum bones of his father and the
other 'specimens' were returned to Greenland. Minik was not
among them.

OUR SILENIC COMPACT

'The ancients,' said Socrates, 'were uncomplicated, and if a certain rock was known for telling the truth, they would listen to it.'

John Michell

'We watched Laz stomping across the Moon Field to his goal, which was a rock to sit on.'

Jack Kerouac

'According to the report, signed by Paul Martin, the Inspector General of Nasa, 517 moon rocks and other so called "astromaterial" samples loaned out by the agency between 1970 – when Apollo missions began to collect them – and 2010 have gone missing or been stolen.'

The Independent, 23.1.12

For some time now, in these winter days, objects have been disappearing from my house. The obvious ones first, no cause for alarm: keys, pens, books. The books I often manage to relocate by a form of schizo radar; they are part of what I am, I speak their words, I navigate by predetermined instructions. Single volumes, as you know, are never more than a chapter in the great stone library of the city. But tweed jackets, favoured cups, letters, knives? It's age, I thought. You need less. A process of editing, winnowing away at the catalogue of nuisance. The head feels lighter. On certain mornings, despite the creaks and grunts, I float free like a ghost, nightsweat sheets sticking to a lively corpse.

I was lying awake, it was around the turn of the tide, the hour of the wolf, of mangy foxes butting lids from rubbish bins, screeching in savage coitus, devouring domestic cats, a massacre of the innocents, when my sleep resistance was

disturbed by a thump in the hallway. One night, returning for the third time from the bathroom, I saw a spectre standing there, featureless face made lurid by the yellow streetlight, but the presence was mute. Absolutely so. It absorbed sound. As I moved towards it, the intruder thickened into shadow: one of our house's previous tenants, returned or revealed in a thread of moonwash. The present thump was more like a small animal going down under a stun gun. Let it be until morning, I decided. One of my missing objects re-manifesting, undoubtedly. Just when I'd settled for being rid of it.

After a poorly attended public reading in an out-of-hours launderette, earlier that week, I had been approached by a man called Norton, who claimed to have written a London novel. The thesis he was promoting – I'm not sure what he wanted me to do about it – was that all vanished things on earth, especially lost or destroyed manuscripts and alchemical formulae, are transported to the moon. This was an old fairy story and of no special interest. Norton shimmered with repressed anger, his heavy head twitched towards you, in a stalled butt, as he muttered and drooled. The flesh was grey as pulped paper: dry, calciferous, crumbling with bone dandruff, creased and printed with the ink of a thousand unrecorded crimes. Shaking his hand was like dipping a paw into the sand tray from a derelict animal shelter.

'There's a rock in a room in Redchurch Street,' he said. 'I could arrange for you to see it. They swear it can perform miracles, raise the dead, develop photographs that have never been taken.'

'Take a card. Send me your contact details.' I blustered.

He tucked the scabby rectangle with the reproduction of the poster for Fritz Lang's *Frau im Mond* under his cap. The address on the back was of a retired bookdealer in Folkestone. I owed him a favour.

Norton believed - and he produced, as these madmen always do, a newspaper cutting in support of his conspiracy theory - that agents of government were combing the city for fragments of American-owned lunar rock. Reports in the media were a standard plant. They used the *Independent* because nobody read it. The Moon cops, moonies, Mormons, licensed to kill or queer, knew that the rock market was bearish, overstocked. More lunar material on the Old Kent Road than there had ever been on that dull local satellite in the sky. *They were planting the stuff.* It was cultivated here, in cellars and retail parks, and deep in the tunnels of decommissioned railways. Half the so-called conceptual artists in Hackney Wick were aliens, paid to divert attention from what was going on behind the blue chipboard fence: lunar harvesting, the grafting of rock-flesh hybrids, messianic cults with white-suit prophets. Moon dust was trading in private clubs and alongside rooftop swimming pools where coke fiends took their substitute hits from electronic interference, the pulse beat of blue screens in the winking windows of empty towers. They were growing moon rocks like melons in a Charles Bronson revenger movie. When you bought a coffee from a person with skin like pumice stone, you were closing in on a covert storage location.

'What if we are the moon's moon?' Norton persisted. 'Those fake Yankee voyages back in the Sixties were a sort of homecoming. Nixon's head in a cartoon bubble like a magic show by Méliès. I've told them a hundred times, the moon is the missing eye of Rooster Cogburn.'

Norton's vanished novel, trashed, junked, thrown away rather than mislaid like the objects in my house, featured a one-legged bigot called Sileen. A man more furious than Norton himself. A corrupted X-ray addict who wanders the

country on a series of meaningless triangulations between
conical earthworks, real and imaginary.

'None of them got it,' the hack told me. 'The lunar
reference in that name, the nod at L-F. Céline, the craziest
writer who ever took on London. Sileen is my Robinson.'

I had already decided, even before being accosted by
Norton, to rescue the situation by a day's walk. The rocks
were out there and they were calling to me, barking like a pack
of randy seals. I rose early, squeezed a grapefruit, swallowed
a cup of gunpowder tea, picked up today's Jiffy bag, the one
that thumped on the doormat in the middle of the night,
and headed off in a random, westerly direction. Honouring
my own ghosts and those of my guide and mentor, Arthur
Machen of Caerleon. His grail, my moon. His paperback
reprints, every one of them, gone from the shelves, flown off
like gulls to a downriver landfill dump.

*Wrenched somehow out of the natural order, I have been
plunged into an incomprehensible chaos where I can make nothing
out, and the more I think about my present situation, the less I can
understand what has become of me.*

That's Rousseau, not Machen. Wrong book in right
place. One text fades, another takes over. Books multiply, if
you don't take active steps to prevent it. Ballard, I've been
told, held regular book barbecues in his Shepperton garden.
I'm too superstitious for that. I bury them and hope they'll
calcify. But there are more, every day, used, mint, coverless,
privately published, handwritten, to replace the missing.

'I could think of no simpler or surer way of carrying out
my plan,' Rousseau wrote, 'than to keep a faithful record of
my solitary walks and the reveries that occupy them, when
I give free rein to my thoughts and let my ideas follow their
natural course, unrestricted and unconfined.'

STONES IN A FLESH WALLET

Now even the squirrels eat muesli, or don't, glutted on daffodil bulbs, margarine cartons, bird seed. In Hackney's Victorian squares and heritaged villas, pampered rodents can afford to spurn the health food option. A ziggurat mound, like vitamin-enriched sawdust, has been set out for the flying rats of the inner suburbs. As I bend forward to stare intently at the paving slabs, the river patterns of crack, the persistence of weeds and mosses, a creature with hooped spine bounds forward with a rippling motion, like one of those concertina sets of metal rings that snake down childhood staircases. All the discrete points through which the rufous animal passes are preserved in the mind's eye, wavelike, serpentine. And this episode only confirms the motif of my walk: a mindless flow against unarguable obstacles, the ancient rocks and stones that hold down the neurotic spread of our city.

Here was my idiot-simple proposition: psychogeology. The beach beneath the pavement. 20,000 streets under the sky. The rocks of the geological collection at the Natural History Museum in Exhibition Road, South Kensington, were calling me in. I would come to them, across London, connecting with, recording, investigating – and listening to, that above all – a chain of glacial erratics, Aberdeen granite lumps, public art boulders, kerbstones, unnecessary cladding, erased memorials, and demolished terraces with the split heads of Coade stone effigies. The lapidary chill in our blood, the lime mortar in our bones, takes sustenance from a chain of volcanic detritus left behind or exposed in public parks. Interventions from who knows where? Deep space? Flying saucers? Pre-history? Blocks of never-completed medieval cathedrals risen from the sinking turf? Overturned rafts, between quarry and construction site, in the days of the river

marshes. Fossils swim through ocean beds of Portland stone churches, encrusting the statues of proud dignitaries. Crude megaliths, pulsing with faint prelapsarian signals, sprawl in shallow declivities. Disregarded by tourists and speeding urban commuters, they bask in the achieved invisibility of things that have always been here with no requirement to explain themselves. Until they are trapped, captured on film, measured, catalogued, and removed to the benevolent reservation in Kensington, where crowds wait in an orderly mob, to be granted privileged access. This chambered ghetto of stolen boulders and displayed crystals was where I was headed on a bosky, blameless morning, under a clear sky and that unwitnessed caul of stars without number. Pebbles take shape from sediment in our waste. Rich red wine, game, sack, maggoty cheeses. Samuel Pepys cannot become a true diarist of London until a barber-surgeon has fished his urethra for the stone he will preserve as a sacred relic. A smooth meteorite, cultivated on the inside, in the heat and flux of a maturing libido, is hooked on a blunt harpoon.

They call it the Snake Park. And promote it, in a tidy enclosure beneath an elevated railway, alongside all the other parks by which London is now defined and divided: retail parks, theme parks, business parks, car parks. Where a decision has been taken, to trim budget by abandoning grass-cutting operations, a post is driven into the ground announcing: 'Wilderness Zone' or 'Nature Reserve'. There are two kinds of wilderness and you do not want to be caught in the wrong one. An approved wilderness will be demarcated by an orange mesh fence. It looks like a few yards of captured meadow, but is spared human invasion. This is the quotation wilderness, a mental conceit; a framed folk memory intended to alleviate

the brutality of permanent building works, holes in the road, strange muddy wounds allowing the curious a glimpse beneath the surface membrane of London, the dull clays and chalks that sprawl in a promiscuous tangle above bubbling, spitting magma. The bad kind of wilderness is also known as 'wasteland', as in: 'The Lower Lea Valley was a wasteland. There was nothing there.'

The Snake Park had been established, for many years, on wasteland. It was enjoyed by generations of infants and those who prey upon infants. Then, like the railway above, it fell from favour. It remained, on its own terms, occasionally visited, frequently padlocked under threat of improvement. My interest today is in the snake itself, the votive beast, a humped and heavy-headed artwork in painted concrete. It lies, glutted, replete, jaws gaping, nodding in the direction I must travel, southwest. Concrete is the slush of stones, the solidifying excreta. Moon rock mixed with paper mania.

A snakestone is an ammonite, the heat-print of a mortal creature translated into some more permanent form. A solid pictograph on which to break your toe. The classic ammonite is coiled like the horn of Ammon, the Egyptian ram god. This municipal Hackney snake has a Mexican or Mayan quality, skin shimmering with tesserae, ceramic and glass fragments implying movement, a slither and stretch to excite the small children who clamber over the long arched back or dive beneath the hungry belly. The decorative surface carries the life force, responding sympathetically to the rattle of passing trains. The stone is slack, cooked up, impure; sand and slurry offering no connection to the geological trail I am trying to establish, my star-path across the covered fields of London.

A lively local, a lady with a small white dog, is in

conversation with an elderly Muslim wearing a sweat-stained
lace skullcap.

'When they get palpitations...' She prods the supine dog
with her stick. 'Heart problems, just like us. They cough.'

She pokes her pet again, aiming for where the troubled
heart might be. The pooch spews up a hairball. And some
loose shred of entrail. He wheezes. But he couldn't be accused
of a cough. *She* coughs, by way of example. He raises a limp
paw. He remembers another life as a monk in Tunisia. Sand
scratching at the mouth of a cave.

The old man, tilting to the west, dragging a stiff leg, eyes
me with suspicion, when, resting on a cold metal bench, I
sketch a few preliminary observations. How sinister they seem,
those brass stirrups hanging from the tree to entice undersize
climbers. How strident the mechanical voice anticipating the
arrival of trains. 'Crystal Palace. Highbury.'

Two of the greys, the Moon Cop agents, are lounging
against the high railway wall, like detached graffiti. Slender
as paint, they are exchanging sepia postcards of long-skirted
tennis girls for a phial of crushed stone worms preserved in
vinegar. I have no doubt that they are tracking me. They have
lunar eyes without pupils. When I look again, they are back
among the brickwork. The vinegar is a piss stain on grass.

Very soon, an aging man in an aging city, I am advancing
into a pattern of street names I inscribed years ago, with
no sense of their meaning, but with an irrational belief that
they were somehow connected. I half-recall lines written
by a stranger, myself, aged thirty-three: 'Reels in the Blake
skein... Orsman Road... Whitmore Road, Hoxton Street, the
Latter-Day Outpourings revival.' *Orsman Road*. The Muslim,
after his ritual conversation with the dog woman – I imagine
it happening every day, same words, same time – has an

established direction of travel, tacking through preoccupied pedestrians to the pencil-thin mosque.

Orsman Road, deserted but carrying an ominous freight of memory, was not my first choice. Permissions, along the canal, were again suspended. More developments, more non-elective futurology. A barrier to which has been attached notices proclaiming the erroneous information that the canalside café with the celebrated coffee was still open for trade. And that its bicycle-repairing neighbour was eager to service punctures or to adjust gear ratios. The print on the laminated notices had run. The businesses in question were shuttered. Nobody was enraged. We encountered the fence, we nodded. This was London. A boy with prominent ribs and dirty pink trainers was recording the smeared notices for his portfolio. He knuckled his groin and snuffled like a peccary.

In my case, on my geomorphic walk to Kensington, the banishment was a blessing, redirecting me to the legendary Orsman Road. I associate this tight tributary, overseen by defunct or revamped warehouses, with shamanic practices, horse sacrifice. Men stitched into reeking horse hide. Mounted skeleton, stone-ribbed 'orseman. That amputated Cockney 'h' triggering thoughts of a severed head, guillotine slicing off the first letter, the veins and threads of a thick equine neck. Godfathering my perverse quest.

Here is a bucket-like container known as a skip. Here is a display of rock and rubble. Broken buildings and dug up pavements are geological by-products returned to their original unexploited form. Odd words, like skip with its Scandinavian origin, older than Ikea, sit comfortably with odd stones. Stones, in all shapes and sizes, split, cut, smashed, are a mute alphabet. They are not telling us anything, they talk

to themselves in their own language. A forgotten alphabet constructed from bursts of signal and flare. One stone to another. Grain to grain. Crystal to crystal. In Trinity College, Dublin, where I had been a student, the term skip was applied to college servants. Dictionaries of etymology suggest an ugly colonialist derivation: 'skip-kennel'. Dog of the slums. Gutter-jumper. Native. Doing what has to be done in a Viking seatown. Let them suck stones in their porridge, the masters. Let them break teeth on pebbles painted to look like eggs.

Behind and above the skip is a warehouse I have visited on several occasions, the reserve collection of the Museum of London; superseded ethnographic and anthropological displays. Horizontal totem poles stacked like lumber in a canalside woodyard. Cabinets of ticketed Aboriginal and Inuit skulls waiting for candles. Whole streets, blitzed in war, or torn down to improve the image of destruction, are stored within the vaults and galleries of this covert building. The greys have colonised it like lunar termites. They polish slivers of meteorite for skin grafts, brain transplants. They orgy with wax fruit and art catalogues.

I thought of a documentary I had glimpsed, the night before, in a Chinese fast-food joint. Parrot fish with their powerful jaws are capable of chewing into rock, munching coral and shitting sand. Those picture postcard atolls with their virgin beaches, beloved of cartoonists, pimped in honeymoon brochures distributed by merchant banks, surveyed by atom bomb tests, are the waste of dredger-mouthed fish. The new restaurant alongside this spoilheap of cargo-cult trade goods and street furniture is offering, as its speciality, 'stone-baked pizza'.

What I am attempting to trace is the line, laid out like a Richard Long sculpture, of granite kerbstones, heavy grey

and sepulchral, cold to the touch, leading me westward, and faithfully shadowed for mile after mile by thick yellow streaks, a sludge of paint as an over-emphatic subtext: you cannot park here. Georges Perec taught us how to read such streets: 'Underneath, just underneath, resuscitate the eocene: the limestone, the marl and the soft chalk, the gypsum... the sands, the rough limestone... lignites, the plastic clay, the hard chalk.'

This area, its obscurity threatened by development, was made from river terrace deposits, after the diverting of the Thames. The zone of Hackney Gravel through which I was now advancing was known to geologists as 'Artificially Modified Ground', a surface 'wholly or partially disturbed by human activity'. Its symbol on the chart was a cloud of bar-code rain, a box stretching from London Fields to the boundary with Islington.

As I approach the Gainsborough Studios, a trim young woman in red passes in the other direction, conducting an intimate, and one-sided, conversation with a tongue-dangling greyhound. The old film studios, now a Germanic courtyard development, are dedicated to, and dominated by, a vast boilerplated head of Alfred Hitchcock. The metal from which this hieratic totem has been assembled stays close to its origins in the earth, the heated stone. Spindly leaf patterns play across the protuberant lower lip and the well-fed cheek flaps of the eminent director. The alien, the fat captain, controller of greys: cut the frame, watch the woman. The scale of the thing, the swollen giganticism by which all cultural references are now made, acts like an iron-cored meteorite on our individual compasses. We warp and bend, absorbing all the back-catalogue film references encrypted within this hidden

courtyard: the vertiginous rear windows, the grassy knoll from *Torn Curtain*, the magnetic pull taking us away from *North by Northwest* to the gravity of the waiting rocks in their cases in Exhibition Road. Was Hitchcock just a head? You had to imagine the rest of the man, the great belly, the belt worn tight to the breastline, buried deep in the clay and silt of a Pleistocene bed.

In Shoreditch Park is an object that balances the Hitchcock head, a freestanding megalith as hard-travelled as the Sarsen stones of Stonehenge. And as mysterious. I recognise the carved intruder as part of a chain of stones, set in parks, and beside housing projects, between Shoreditch and the Thames at Shadwell. You could see them as glacial erratics abandoned by the retreating ice. Or as follies laid out, at enormous expense, for a corporate project whose purpose has never been revealed and which is now quite forgotten. This igneous rock was the immediate inspiration for my walk to Kensington. It shines with discriminations of feldspar, quartz and mica. And is dressed with manmade holes which give it a cartoonish anthropomorphic quality.

Who would leave such a thing in such a place? The natives, sweeping through on their preoccupied trajectories, ignore its immense and shamed bulk, as they scamper to secure a Boris bike from the set of hitching posts. The stone has been labelled by John Frankland, the artist responsible, as BOULDER. But if you think that word suggests a random obstacle, the Shoreditch Park erratic is no such thing: it's *intended*, blasted from a Cornish quarry, heaved, manhandled, hoisted and transported like a chained captive to its secular resting point in a public park. The holes you might take for sculptural embellishments or Bronze Age workings are nothing of the sort. They are aids for climbers. The rock is

here for urban recreation: 100 tonnes, four metres high, and soliciting engagement. A tricky ascent, graded V4 in the advisory booklet, is known as 'Inner City Riots', a prophetic stab in the dark. Athletes travelling to Shoreditch to take up the challenge are advised not to leave any valuables in their parked cars. They will arrive, as if in the Lake District or North Wales, in a motor vehicle. And they will be reminded, in the brochure, of where they are. A criminous backwater where potential upheavals can be conquered by crabbing over a standing stone. 'You are in London,' says the artist, addressing his boulder as much as the casual visitor, 'and not in Cornwall.'

A marble bench from which to view the rock has been embossed with the words NEWTON STREET, in homage to William Blake's naked youth, stooped over his compasses, against the teeth of a stone circle. Shoreditch Park, today, is a pure geometry of shadow-lines (human, light pole, tree) making alignment with the stolen boulder. Patches of black lichen are like smears of squashed bug, dried heart-blood sweated from deep within the stone. Lesser rocks, chipped or carved, surround the monolith in an unconvinced scatter.

The Newtonian compasses have been primed, so I will follow their dictation to that minatory workshop on Shepherdess Road, H. Bestimt & Co, Feather Merchants. Dark stairs at the rear, rooms let out to single gentlemen of a retiring disposition. I came here, once, with my young son, to deliver a book. He never forgot it, the reek of scorched feathers, plucked birds, wings pegged out to dry. A fine dust waiting to lodge in the cracks of your skin, the corners of your eyes. If the chain of orphaned rocks had no weight and floated down on the landscape like loaves of Magritte bread, then the feathers were as heavy as steel. Plumage House, they call it.

The small park, on the other side of the road, residue

of demolition, felt like a burial ground. It contained a group
of granite coprolites, sculptural interventions in the form of
carved sheep. Spiral horns butting out of stone. I noticed a
walker come away from the path, to micturate like a horse in
the sheltering bushes. Like Céline in the Seine. Trail-marking,
I suppose, for future reference. Vegetation, dense and tangled,
was pierced by spokes of weak sunlight. The man, duty done,
stood unmoving as I walked away. A German documentarist
described the risks of being drawn so deeply into the act of
gazing at a jungle. 'Stay still for five minutes and the plants
will eat you.' Keep moving, stone to stone, and you become
the thing that you imagine. Already, I sense the clots of lichen
forming on my fingernails. The pissing man, grey raincoat,
hands in pockets, male member dangling and dripping, waits
there, among the shrubs, to be noticed.

Even the paper hoardings on City Road, near the inlet of
the City Road Basin, reference geology as a selling point, a
style issue. 'Designer kitchen with granite work tops', so they
assert, will feature in every virtual flat in this photographed
but unbuilt canalside development. You can see, looking
across the water, that not only does Hawksmoor's obelisk at
St Luke's, Old Street, stand on its relationship with the shape
of the narrowboat basin, but it predicts an alignment with the
tower of St Mary's Church in Upper Street, Islington. The
advertising hoarding is no help, suggesting, as it does, that we
are in an occulted city under the laws of the voodoo of capital.
Estate agents hold the key but it is only available to bona
fide investors. A rectangle has been depicted as a labyrinth
of unreadable black shapes: 'A Bar Code app is required to
read the above Code'. To add to the confusion, my geological
chart warns of shifting gravel beds, areas in a sour yellow, like

quicksand, and incense-bearing clouds of ecclesiastic purple, the silts of Enfield and the gravel of Finsbury and Boyn Hill.

Time for a coffee hit and the Jiffy bag.

HEADJOB

My spectacles were smeared, where they sat on the table, beside the empty coffee cup and the saucer of grainy raisin crumbs. The weak resolution image of the busy breakfast bar was fit for purpose. You can't tell the time of day by the food on offer. I didn't know what to make of the bulge of assorted papers, the ant's-footprint script, the bendy X-rays. The author of this mess, inflicted on me in the middle of the night (the only postal service currently available, delivery-by-hand madmen), was the person Norton mentioned, Sileen. Or, more probably, the whole episode was a conceit by Norton himself, mythologizing a myth. Or, Norton had found the papers in the street, liberated them from an out-patients' clinic, and made them his own by virtue of selection. Now, I suppose, they were mine. My problem. Or yours. If you have come this far.

What I could make out, among the diversions, anathemas, lists, quotations, thefts, was that Sileen was sick, fatally, terminally: a pumpkin tumour. The map of his brain, scanned in some humming magnetic torpedo, revealed a growth like a coral reef or meteorite. The inference being that the lump, the intruder, was the essence of the man. He was carrying one of the missing samples of rock, stolen from the moon by crass astral tourists, in his head. Around that germ, a rind of personality had formed: Sileen. Now the rock was fattening, flourishing, glossing. Until it would split the host skull. And dictate its terms of surrender.

Meanwhile, the victim, in his vanity, believed that

scribbling enough words, pouring out his incoherent
confessions, would hold back the progress of the parasitical
malignancy. He would walk the shape of the tumour, the
phantom brain pregnancy, into the labyrinth of London. The
X-ray of Sileen's skull was laid over a map of Clerkenwell, over
the westward stepping route that I was tracing. Towards the
colony of rocks in Kensington.

'If I can reproduce the outline of my disease within those
streets where the prophet Emanuel Swedenborg once lodged,
and fed, I will live, the thing will go into remission,' Sileen
wrote. 'But it must be done with precision and within a single
day. We are bugs chasing a line of sugar.'

'Do you mind?'

The grey at the next table put down his newspaper
and reached across for my cup. Nothing left in it but a scum
of brown froth. He licked. He uncoiled a lizard tongue and
scoured every last trace. Then, as I made for the door, he
tipped the crumbs from my saucer into an evidence bag, a
sachet he pressed to the thread of his lips, before returning
once again to his Mexican newspaper.

NEWLY DISCOVERED DOCUMENTS FROM AN
EXPEDITION THROUGH THE NETHERLANDS OF
THE LOWER LEA VALLEY IN SEARCH OF RUMOURS
OF A RIVER OF SUPERCELESTIAL STONES AND
THEIR EFFECT ON THE BEHAVIOUR OF THOSE
STILL LIVING WITHIN THIS TERRITORY UNDER THE
SHADOW OF THE MAYAN ENDGAME PROPHECY

Closing on the river, after a diagonal traverse through a zone of abandoned hospitals and neighbourhood shops with fire-blackened boards over their windows, we emerge on a cropped meadow marked with a series of complicated geometric figures. The markings, it appears, have been made with some form of birdlime or crushed bone. The Cornish menhir, when Bosch tries to make a record of its position, height, surface features, gives off such powerful radioactive waves that the photographer believes his plate has been ruined. He pulls down his soft hat and puts on a pair of smoked-glass spectacles. 'This is not what I expected,' he says.

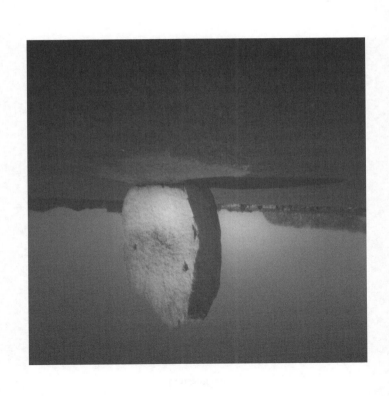

An hour later, after numerous enforced detours around unmanned checkpoints, but always with that sense of hidden eyes watching us from the woods and ivy-draped guard towers, we discover a decommissioned sewage treatment plant. A field of sacrifice has been prepared, combining a star-chart of imported standing stones with concrete slabs and stained altars. Holes have been punched in some of the blocks. And one stone, reminding me of a line by the poet Charles Olson, features an incised letter E.

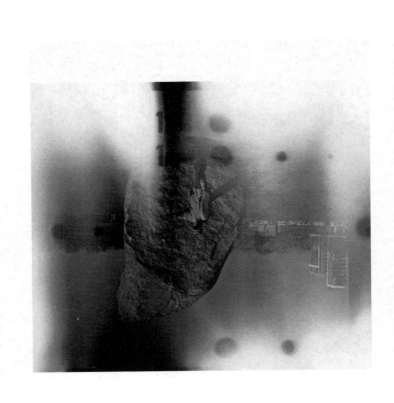

TORTOISE RESCUE

'Like the man who had to pass the same way twice to cast a shadow.'

Jean Baudrillard

At the crossroads of the Angel, stone is cladding, a quotation. I found myself, in a sentimental gesture, echoing Scott Fitzgerald. The way that walkers, from time to time, need to touch and stroke and pet the stone, graze a knuckle. 'He felt suddenly of the texture of his own coat,' Fitzgerald said, 'and pressed his thumb against the granite of the building by his side.' The angels of Islington are banished, landlocked: wings of granite, limbs of marble. Loose Grecian wraps around unsexed goddesses who perch, flightless, on plinths in shady parks, under moulting London plane trees.

Housmans Bookshop is a necessary pit stop before the calamity of the great metropolitan stations, cliffs of brick under permanent revision. King's Cross was formerly a site of memory, the war dead, fire victims, those killed in bomb blasts: names engraved on marble. Much of this material has now been sent away, into storage in Acton. The wrong geology for an age of pastiche. I pick up a book, on grounds of size, weight (lack of), and title: *An English Figure (Two essays on the work of John Michell)*. I felt that this might be the right moment to reconnect with the author of *The Old Stones of Land's End*. Michell, whatever is said of him, did not feel the need to blast lumps out of Cornwall and to transport them back to Notting Hill to prove a thesis.

The British Library, relocated from its established nest in the British Museum in Bloomsbury, has taken some account of its

position on our geological track. Sands and gravels, London clay, they are exposed in the perpetual trenches dug by utility companies. The approach to the Library is a provocative confusion of art and artifice, the sacred and the profane. A sunken amphitheatre, dressed with stones on plinths, is a site of authentic attraction for urban dreamers, coin-tossing gamblers, resting pilgrims. Many of whom will advance no further towards the doors of the Library. This forum, with its sculptural interventions, open-air café and hedged enclosures, is enough: a retreat trading on its proximity to the walls of books, the multiple voices at play against the contrived tranquillity of a place of passage.

The local sculptor, Antony Gormley, a hierophant of grand corporate developments, has given the small amphitheatre (a teasing recollection of the Irish and Scottish drinking schools who met in a declivity one of the survivors, John Healy, called 'The Grass Arena') some necessary gravitas. Eight rocks assigned to eight planets encircle the sprawled visitors. Pressed into the stone, or emerging from it, are hands and limbs. Like prints of the future dead. Or stopped citizens of a city overwhelmed by a volcanic eruption, bathing in hot lava, and lingering ever afterwards as images fixed in the cooling grain of igneous rock. These sophisticated trophies, floating between the status of museum-quality geological specimens and a prescient anticipation of coming catastrophe, have an evident relationship with the humbler sheep-stones of the little park, off Shepherdess Walk. The art aspect rubs off, the rocks hold their position on the flight path of a potential drift through London, leading inexorably towards Exhibition Road. That culture-dormitory in Kensington where the regiment of greys foregather, besuited or naked, rocks held in their hands like eyes taken from the skulls of the unborn.

An Irish writer who always kept pebbles in his pockets, flattened schist, ovoid serpentine, fled disgrace (after soliciting it for so many years) by enduring a downriver exile in Silvertown. He saw his nocturnal train ride, overground, then under, to South Kensington, as a stretching of that precious silver, by rail, towards the silent rocks in their glass cabinets. Subterranean snail tracks. The total anonymity of his position as a night watchman rhymed with his notion of what a writer's life should be. One of the senior curators, so he told me, cycled through the silent galleries, when the crowds had gone. He held loud conversations, in Latin, Greek, old Cornish, Norse, as was appropriate, with favoured stones. When, inevitably, the dusty museum was challenged to justify itself and to pay its way, they dedicated a chunk of the budget to art promotions. A troop of nude figures, white, plastered in gypsum, living plaster casts, statues liberated from their perches, roamed the corridors. The writer took his cards and returned to Dublin.

Dominating the British Library terrace, and facing firmly east, is Eduardo Paolozzi's three-dimensional rendering of Blake's *Newton*. In the lee of the Novotel, the bronze giant stoops to his compasses, like an unclothed supplicant attending to a football coupon – which is appropriate, the artwork is sponsored by Vernons, Littlewoods and Zetters, the pools conglomerate. As at Shoreditch Park, the combination of boilerplated figure and captured rock (in this case the necklace of small Gormley boulders) confirms the significance of the site. But, if I want to make the transit to the sphere of influence of Kensington, away from this reef of railway stations and the approaching concrete launch-ramp of the Westway, I will have to attempt a forced detour into older gravel deposits from the Pre-Anglian epoch. Marylebone

represented a surface shift, less granite, more marble. Sleeker surfaces dressed with blue plaques, memory prompts no pedestrians trouble to read.

At what point, precisely, do the stones of Kensington replace the boulders of the East London gravel beds as the dominant influence, the force warping my magnetic compass? Not yet, not here, where the pressure is off and dignified grey buildings are protected by discreet security, winking cameras, window boxes, and plump chauffeurs waiting in black cars outside cancer clinics.

In Woburn Place, right opposite the plaque for WB Yeats, is a newsagent's shop with a copy of *Frankfurter Allgemeine* on display. The front page spread, today, features a black and white photograph of a naked man – Newton released from his bronze carapace – posing against an insecure heap of boulders. I have no idea what the story is about, but it confirms my direction of travel: ignorance. The creamy limestone of the middle Jurassic age, quarried from Cotswold hills, is drawing me, unconsciously, to the west. Like a poultice against infection. Against the residual West End fear of a wind from the east, bringing poverty and disease.

Portland Place, with its BBC bastion and its reservations of tame architects, is also Portland Stone. Eric Gill chopping the male member to dimensions acceptable to Lord Reith. The muscular Edward Bainbridge Copnall, shirt off, beret on, takes his chisel to an hieratic relief figure symbolising *Architectural Aspiration* on the façade of the RIBA building. There is now, distinctly, as much geology above ground as below: the thin granite margin of the kerb, the cut blocks from quarries on Portland Bill (decorated with Jurassic oysters), paving slabs of Mansfield Stone (from Nottingham

and the Permian age), sandstone slabs from the suburbs of Edinburgh. A catalogue as eccentric as the seething human shoals competing for space on pavements and in gutters. There are slumbering, flowing, deeper layers of clay, sand, chalk and shallows of drift-geology exposed by those casual archaeologists of morning, Irishmen with picks and shovels, drills and dredgers. Spoiled priests of London's subterranea.

An agency grey, posing as a rep for some Saturnian religious cult, hands me a card. TURN BACK. FROM WICKEDNESS. AND THE SINS OF THE FLESH. Get with the programme. Tithe us your lifeblood. His tongue flicks like a whip, snaffling a couple of mating flies.

Drifting with the faint southwesterly pull of the Kensington reservation, the sliced rock samples in their display cases, I find myself on Weymouth Street. STONE HOUSE announces a brass plaque, being polished at this very moment, by a man in a brown suit. An elegant, stepped, flower-bedizened enterprise, like a faux-Thirties hotel from an Agatha Christie television adaptation, dealing, behind the intimidating foyer, with some form of biotech research and sinister marketing. I thought: vivisection. I thought: John Cowper Powys. I thought: *Weymouth Sands*. Powys, of all English writers, loved his stones, the animist philosophy, the unconscious surge of life to be found in reaching for and touching a shard on his track. Preoccupied, alarmed and excited, by the imagined tortures and perversions of science within protected buildings, Powys ritualised a counter current by fixing 'one of those geographical points on the surface of the planet that would surely rush into his mind when he came to die'. A bench, a bent nail, a lichen blot, they contained: 'the concentrated essence of all that life meant!'

Property developers, not much interested in geology, have issued their own bulletin on Weymouth Street, from what might be described as an anti-Powys perspective. They see this seductive tributary as consisting of '89 properties, 78 homes'. The average 'asking price' – you asking? – is £1,550,924. The great advantage of the street, so the promoters suggest, is that the nearest petrol station is only ten minutes away. There is a convenient Orthopaedic Hospital on Bolsover Street, but you will have to endure the noise of ambulances.

Manchester Street and Manchester Square remind me of the curious psychogeographical doodlings of the Spitalfields hermit, David Rodinsky. Rodinsky annotated his friable copy of the *London A-Z*. One of his routes carried him to this point. I walked, as nearly as I could, down the fading biro track, but never established a motive for his projected (or remembered) journey. I was tempted to inspect the Wallace Collection, standing proud to the tree-enclosed square, for paintings that featured significant rocks. But the presence of a young European woman, so much a quotation from an artwork I failed to identify, long legs and country-splashed Wellington boots, threw me off balance. And I photographed, in compensation, a ram's head embellishment in Bath Stone. The poet Ed Dorn, lodging here in the mid-Seventies, was captured on camera in the square gardens, alongside Marianne Faithfull and the rest of the Dunbar family tribe. He reports, with relish, that a key to this enclosure costs two guineas: 'that denomination a remoter clue than it once was'.

As we turn back, eager to follow a murky tributary to the broad stream down which we ventured into this oppressive terrain, we recognise the influence of the Nazi architect (and prison walker) Albert Speer: The stones are sharp-edged, Vorticist, minatory; a matter of steps and slits and obstacles. Machine-gun nests and tank barriers from previous wars, designed to hold off water-born invaders, sit alongside derelict modernist towers and stumpy obelisks. The elevated footpath has been treated with hardcore, gravel, tarmac, until it resembles an insane seven-mile airstrip for some Howard Hughes flying machine that will never get off the ground. A mad-eyed shepherd, dragging a dog on a string, warns us that we risk immediate arrest – and disappearance – if we persist in using our cameras. The strange objects on the roof of the apparently deserted tower block are surface-to-air missiles.

Escaping into streets as close-packed and busy as the outreach settlements, shanty towns and favelas of São Paulo, Istanbul, Mumbai, we register the presence of monoliths, granite boulders, left on wide pavements or positioned alongside blocks of cheap public housing. Some of the stones pay tribute to drowned nuns and forgotten poets. Others are naked of inscription or obvious purpose. It is clear to me that these arcane and perverse acts of geological manipulation have a political purpose. Bosch, confused by the way that some of his photographic plates are under attack from electromagnetic rays emanating from the stones, while others are clean and neutral, insists that these manifestations are entirely natural. Either the rocks have emerged from the ground, as with those cut medieval blocks, intended for Ely Cathedral, rising hundreds of years later, from drained fenland – or they have fallen from the heavens. 'Observe,' he says, 'the scorch marks, the surface damage. The trauma.'

MORPHOLOGY OF A SOLITARY CORAL

Marble Arch was such an obvious portal on this expedition that I was tempted to leave it out of my narrative; but passing through, as one might have passed, by John Nash's original intention, into Buckingham Palace, opens a direct route to Exhibition Road. The geological gravity – and the gilded frivolity – of the Albert Memorial and the Albert Hall hurry my tread. Hyde Park wraps walkers in meadowland, aborting predetermined plans, revising destinations. Excursionists wander in delirious circles, before giving it all up and flopping on the grass. Where you expect agency greys, coupling after their fashion, or watching with blind eyes from behind royal trees, they are not to be found. They have vanished.

The Serpentine Gallery triggered a memory flash of the poet and performance artist Brian Catling, in residence, circumnavigating the paths and pavilions before tapping neurotically at the glass, a pebble in his mouth, his eyes bathed with thimbles of ink. Inside the current gallery was a photographic print of some Hackney flats by Tom Hunter, a mirror image of my starting point, letting me know that my journey was almost done.

The queue on Cromwell Road, outside the Natural History Museum, is substantial. And growing by the minute. Has there been a sudden convulsive urge to visit the stones? What looks like another transported monolith, peaty, iron-enhanced, and erected to stand against the artificial towers of Alfred Waterhouse's Romanesque depository, is actually a tree, 'older than the dinosaurs'. A fabulous stump extracted from rocks with a birth certificate going back 330 million years. This withered tusk, exposed like a Tyburn sacrifice, outranked the Victorian museum, where a crowd of

the indifferently curious clustered and mobbed.

The only viable strategy was a retreat to Exhibition Road, now fenced, excavated, brutalised. Teams of sponsored improvers were 'Transforming London's Cultural Heartland', by undertaking the sort of activities that deny a zone its geological pedigree. Visible rebranding, by way of clawing away the surface, trenching, scraping, erecting granite bollards and avenues of orange cones, distorts the older identity, the long-established pattern of flow and access. London Clay, from the ocean bed of 50 million years ago, is easy to tunnel. Tiled passages run away to other collections and educative gatherings from the era of empire. Sound reverberates from packs of children enduring cultural improvement with varying degrees of grace and acceptance.

If you come in at the door closest to the former Geological Museum, you bypass the restricted admittance policy of Cromwell Road. JE Gray, Keeper of Zoology (1840-1874), believed that exposure to this overwhelming abundance of specimens and categories, the ever-expanding treasures and taxonomies of the museum, drove curators and keepers mad. The vast building turned humans into hungry ghosts. And ghosts into stones. It wasn't the room of geological maps and charts, the one with no official opening hours, that I was seeking. Nor yet the exhibitions of sea lilies and aragonite. Nor moon dust scratched from the blind eye of our nearest neighbour. Something more potent was drawing me in, a density of reference more convoluted and complex than anything encountered on my walk. Something unearthly but utterly present: *a meteorite.*

Admiral Peary dragged, at great cost, and after several fruitless attempts, just such a brute, thrumming with ferrous memories, from Greenland to New York. With this difference:

he chased rumour, snow weapons tipped with material
grudgingly excavated from this extraterrestrial boulder. The
Cranbourne Meteorite of Exhibition Road appears to have
been here forever, from before the time when the first brick
was laid; so that the London of the walkers, the ones whose
peering faces were now reflected on the dusty glass of the
protective display case, retreated, layer by layer, from the
killing indifference of the captured specimen. Here was a
tumour we could never extract, older and more savage than the
London Stone from which all distances were once measured.
The surface of the meteorite was indented with pressings and
finger marks, as if earlier guardians had tried to claw their
way inside, to reveal a locked secret. Rocks fly, humans crawl.
Out in Stratford, the furthest easterly point on my necklace of
accidental boulders, was an anonymously crafted memorial,
about the size of the meteorite, dedicated to Gerard Manley
Hopkins. That small boulder, or grey lump, unlike this one in
Kensington, has a message on an embossed panel. A fragment
of text. The meteorite sucks words in, solicits narrative, and
offers nothing in return. The Hopkins rock pays tribute to a
drowning, *The Wreck of the Deutschland*. Five Franciscan nuns
lost at sea.

> *Rhine refused them. Thames would ruin them,*
> *Surf, snow, river and earth*
> *Gnarled: but thou art above, thou Orion of light.*

A CONSIDERATION OF THE RECENT RIOTS

After Sileen's death in a narrow room in a warren of such
rooms, most of them deserted, in a penitential building under
revision, close to his beloved London Hospital in Whitechapel,

Norton inherited the papers. The archive. A cheesy quilt of padded Jiffy bags, stuffed to bursting, from which the amputee assembled a primitive sleeping pouch for his post-procedural hibernations. The X-rays to which he was addicted, and for which he had to invent (and justify) an even more elaborate pathology, finally did him in. The fictionalised cancers obeyed his narrative and became real. A once resilient slab of liver, the seat of his soul, metamorphosed to slate. On it, his alcoholic excesses scratched a rudimentary alphabet. The hairless scalp, pocked and pustuled with burns and festering scabs, slopped as he walked: a drowning cat in a bucket. The bone interior calcified, squeezing the brain pip like a pulpy orange, until there was nothing there but an impacted nodule of alien rock, moon dust. Sileen: Lunatik. A sorry meniscus, a scythe of impure light in the dirty window, doubled the motif on the summit of the mosque's pencil tower.

Norton, more dealer than writer, assessed the word hoard with a cool eye. Nothing to be done, empirically. Not a florin to be made from these paranoid ramblings. Nothing to steal and rebrand. William Burroughs, Norton recalled, had realigned the energies of London by playing back random recordings in unlikely places. He revoiced the voiceless, the angry fellaheen. What if, delivering by hand from an overstocked rucksack, Norton should post a bag of Sileen's words through the letterboxes of notable wordsmiths, hacks, rivals? He thought of Alan Moore in Northampton (and that reservoir of deep dreaming in the asylum where John Clare, Lucia Joyce and Dusty Springfield blessed the ground with their solitary pains and visions). China Miéville (whoever and wherever he was). Nicholas Royle with his taste for doubles and re-edited ghost cinemas. He thought of public intellectuals like Will Self, who wrote, as he cycled and walked, outwards

from his address on the other side of the river. But Norton felt, on consideration, that this might be a matter of coals to Newcastle. The psychopathology of that last Self book, the slow dissolve of ego, plodding over a crumbling coastline north of Hull, probably came from Sileen in the first place. Norton could put himself around, a leech at debates and readings, a swallower of complimentary wine in plastic cups. Easier perhaps to stick to his own patch and to land this stuff on the locally notorious, those hungry-for-copy scribes and scribblers, the electively displaced of Hoxton and Hackney and Leytonstone.

So when the second bag arrived on my mat, with a covering note from Norton, blaming me for Sileen's tragic demise, and snarling in green ink about my failure to duplicate, footstep by footstep, the route outlined on the X-ray, I was almost ready to make a further expedition. There was no other work on hand. Paying publishers had cut off non-paying authors who had been on the books, as liabilities, for too many seasons. Why not accept the dictation of a dead man? Sileen's diary listed a meeting with an artist called Tapir Knight in a warehouse in Bethnal Green that was about to be converted, under the guise of 'student accommodation', into another termite hutch of expensive flatlets with vestigial balconies. After Knight's name, Sileen had scrawled: 'MK'. Mark? Monkey? Maniac? Don? The accompanying X-ray clarified the situation. A space shot. A lift from Nasa files. A smudge of grey ectoplasm against a grey background. A time-coded capture from a boundless ocean where there is no time. MK: Meteor Keeper. Curator of solar debris. Artist. Guardian of a 4.5 billion-year-old rock. 'The largest meteorite in private hands in the UK.'

Tapir Knight, well named and long-snouted, was naked, ivory pale, and strapped to his unrocking chair with a complex system of bandages, mummy wrappings that left one hand free to move and work.

'Beta blockers,' he said. 'I've got my heartbeat down to twenty-eight. But I can improve on that. One breath a minute.'

He carved, he peeled stone. A fine butcher of rock fruits heated in crucibles. The intonation was Brummie. But Knight chorused his deep and enduring love of this place, vanishing London.

'I never start work before three in the morning, the birthing hour, when there is least tremble from traffic.'

Invited, after being gloved in white cotton, to handle one of the slices of meteorite, not yet engraved with Knight's upbeat messages, his L. Ron Hubbard platitudes, I felt a power in my hand that drove it towards the table, the floor, the centre of the earth. One splinter weighed as much as Burnley.

'I'll never get over the loss,' he said. 'They cut through three padlocks, a metal door and a steel cage. The more security you put up, the more determined they become. CCTV is the best advertisement for a thief. They know there's something worth nicking.'

It seems that kids from the flats, 'darkies' according to Knight's quaint phraseology, had ramraided the exterior door with motorbikes, roared down the corridor, into the lift. They'd come away with nothing but a chunk of rock, which it must have taken six of them to lift. Knight had no television, no sound, no landline. Any interruption would be fatal to the engraving process. He was determined to complete his self-imposed task: to fit Hubbard's ten commandments to slivers of the oldest rock in London. The defunct billionaire prophet with the WC Fields waddle and the jaunty nautical cap told

his followers that they had got time back to front. Futurology, science fiction: they travelled in the wrong direction. What the hacks described as lying ahead of us was precisely *what we had already endured in some remote age.*

'The meteorite,' Knight pronounced, 'is a lumpen disk, a record of a protoplanetary star sequence, a history going back 4567 million years, at least. Necklaces from iron meteorites will far outlive the dust of the skeleton kings who wear them. When our brief earth destroys itself, my fragments of circumstellar debris will carry their inscriptions into the plenum of eternity.'

Around the walls of the studio were photographs of detached crumbs of meteorite enlarged to look like mountain ranges, underwater reefs, icebergs, floating islands.

'I harvested every one of those after those riots,' Knight said. 'Every image has a map on the reverse, the location where I picked up the stones used to attack shop windows and police cars. The riots were nothing more than a psychotic reflex triggered by possession of my stolen meteorite. Disaffected youths, fired by electronic communications, became utterly deranged and started to fling their missiles back into space. Hoping they would somehow reunite into a single block. They were reconnecting with their forefathers, the hut-dwellers and spear-fishermen of the marshes.'

Walking home from Knight's studio, with all those images, those rescued fragments, still playing in the sump of my crocodile brain, I fetched up against an obstacle that I recognised as a public artwork. The collaboration between promoter and commissioned artisan achieves a spasm of political correctness, a display of generic futility waiting for years of neglect to mulch it down to a condition

where it might be worthy of a second glance.

But in Weavers Field, as so often with these interventions, local school kids have been brought in on the act. They have been invited to forge upbeat slogans, which will be baked into interlocking bricks. The warm red rectangles form a chorus of voices around a spindly sculptural dance symbolising everything lost and implied in the park's title. GIVE US MORE TO LIVE FOR THAN WAR. NEVER DOUBT THAT A SMALL GROUP OF THOUGHTFUL COMMITTED CITIZENS CAN CHANGE THE WORLD INDEED IT'S THE ONLY THING THAT EVER DOES.

I had been talking about bricks with Tapir Knight.

'Meteorites are not rocks or bricks,' he snapped. 'They are iron. Men have forged weapons from them since time began: axe heads, knives, swords. And now urban projectiles.' Before I left the studio, Knight showed me a box filled with fossil-enhanced stones, their edges chipped and nibbled.

'Everything in the new computer-generated London denies stone, mistreats wood, spurns the sloppy mud and silt of poisoned creeks. These fools wouldn't recognise a meteorite if it dropped, slap, in the middle of the Olympic stadium. But the stones are beginning to sing. Thick lips of kerbs exert a powerful magnetism, twisting the blade of the compass, like that moment in *Moby-Dick* when lightning strikes and the ship sails *against* the direction of its fallible instruments. We are coming back to the meteorite within ourselves, the ice memories, minerals in the blood.'

The fretful artist, twisting against his bandaged restraint, spoke of bricks made from the residue of the Beckton Sewage Works, human waste coming back, as at the start of everything, to shape the walls that contain us.

I remembered my own fetish for picking out tide-

smoothed pieces of brick on the Thames foreshore, in Wapping or East Tilbury, bricks with fragments of lettering, broken alphabets. And, like a child again, making sentences from the traces of vanished architecture.

All of which is a long way around to arrive in the streets of Hackney at the time of the recent riots. Tapir Knight is so deeply embedded in the matter of place, the mapping, recording, celebrating, that he felt obliged, as a moral duty, to make witness in some way. Not just to the theft from his studio, the loss he had suffered, but to the pattern of meteorite darts and projectiles spread across the borough. You could see in news reports such a relish for apocalypse. A delirious acknowledgement that the fires and trashed cars and supermarket sweeps were the perfect pre-Olympic promo. There was no need for further reportage, images were swallowing images in a self-cannibalising chain. Confused souls parachuted into these war zones stood stiffly at their posts, while all around them danced the children of the area, tweeting and twittering and moving the action on. The flames seemed to run, by some malicious instinct, right down the new Overground Line, from Dalston Junction to the terminus of a long-established furniture store in Croydon.

Knight's methodology was to gather the meteorite windfall, by pram and bicycle. To tour the aftermath of the riot zone, in Clarence Road and Mare Street, and by broader trawls through the territory, like an urban archaeologist; evaluating axe heads and grapeshot, picking up chunks of rock that had bounced off police cars, feeling tender surfaces for damage. He did not ravish the moment with his camera. He transported the fragments, as relics, to his studio, where they could be afforded the dignity of forensic examination

and logging by high-definition lens. At which point, the membrane of the ferrous material is revealed as a landscape, a lunar surface seething with microscopic life, lichen clusters, stains and fissures. That which was thrown is now calmed, assessed, evaluated. The history of small wars is told through a consideration of the spent bullets and shells recovered from the battlefield. Every one of these shards has a narrative. Some of them were recovered by Knight, after he had noted, from his laptop screen, the trajectory of flight picked up in a news report. Sections of garden wall, broken flag stones, they were pressed into service at this instant of social upheaval. The photographs restore gravitas, a stoic cataloguing of chaos. They are a truer portrait of the crowd broken down into individuals for judgement and retribution than that ugly parade of mugshots in the tabloids.

Cities can be mapped by missing cobblestones: Paris in '68, London at the burning of Newgate Prison, Budapest, Belfast. Streets are dug up in reverse archaeology. The stones redistribute themselves, flying through the air in the direction of Plexiglas shields and visored helmets. If you can't trust the digital captures of women leaping from flaming buildings or street actors brandishing their swag, the bricks in a box are hard evidence, pure and unsponsored. Surface drama is bled from the story, so that we can witness the purity of abstraction. We can be grateful to Tapir Knight for finding a way to record this event, modestly, carefully, without hysteria, or any boast that he alone has the solution. No recipes, only rocks. And beyond the rocks, the strange fruit of our most ancient and sacred ritual tools, the chippings of meteorite.

'There is one well-documented account of the use of iron meteorites as a primary source for a hunting technology,' Knight said. 'The working of the Cape York meteorite into

knife blades by the Inuit of Greenland. And, before them, the older beings, the ones who came here from the stars.'

John Minton's 1951 cover drawing for Roland Camberton's Hackney novel, *Rain on the Pavements*, takes the same high-angle view of Mare Street now offered by the helicopter-eye coverage of the recent riots. But there is a striking difference. Minton's aggrieved marchers, holding up political placards, are heading for the right place in which to lodge their protest, the Town Hall. The August consumerists head for JD Sports, betting shops, and stores offering white goods, plasma screens, handbags. Camberton's ambition, like that of Tapir Knight, was to *know* every stick and stone of the borough. The 2011, pre-Olympic flash mobs, like the promoters and salaried bureaucrats and offshore developers, wanted to tear the stones down, to remake the world as a heap of future ruins, a cave of glittering trophies. Knight's photographic portraits, grey and self-contained, are a telling record of this historic moment. Having once handled the chill and the extraordinary density of a meteorite fragment, in an act of unholy communion, they will be back. Our local Inuit. The dispossessed. And next time the city *will* fall. When the vandals decide to remove the London Stone from its nest beneath the paving stones of Cannon Street, the entire fabric of the labyrinth will tear itself apart. 'We are,' as Sileen concluded, 'never more than an extension of our geology.'

Do you feel it yet? The incipient migraine, the clotting, the terrible weight when you try to lift your head from the pillow? Do you sense rock forming within the bone cup? It is coming. It is coming. It is here. I remember the words of Philip Lamantia who remembered, in his turn, the poet John

Hoffman, who vanished into Mexico. Into the mouth of a volcano, as some thought. Into a Guadalajara hospital. 'By night the deltas of the moonspilled planet/ are stoned under his wriggling light. *By day, he chokes the sun.*'

We arrive, as light fades, at a wayside shrine. The relic of some uncatalogued stone-worshipping religion. A granite lump, hewn into a likeness of the area's mythic Elephant Man, has been set in the alcove of a demolished church. It honours: One Unknown Yet Well-Known. The paradox is teasing. The mouth of the crudely sculpted caput holds the secret of place. Wind whistles through the narrow aperture of the stone flute, raising the spirits of the shamed dead. Bosch, putting his eye to the hole, says that he has gained access to a primitive, but pure, form of cinema. The pursued mouth acts like a pinhole. Through it, he can witness remote events in terrifying detail. He can see the shaped meteorite on the lid of an ancient sarcophagus. And, under the trees, where shrouded strangers are whispering their fictions, he captures every word they utter. White mist had risen from the meadows outside, and we watched in silence as it crept slowly into the church porch, a rippling vapour rolling forward at ground level and gradually spreading over the entire stone floor, becoming deeper and denser and rising visibly higher, until we ourselves emerged from it only above the waist and it seemed about to stifle us.